Sarah Naiyer and Khalid Hussain Bhat
Editors

Transcriptomics and Their Importance

nova
science publishers
www.novapublishers.com

DOI: https://doi.org/10.52305/IZBB7081

NOTICE TO THE READER

Library of Congress Cataloging-in-Publication Data

Names: Naiyer, Sarah, editor. | Bhat, Khalid Hussain, editor.
Title: Transcriptomics and their importance / Sarah Naiyer, PhD (editor), Research Scientist, Department of Microbiology and Immunology, University of Illinois at Chicago, USA, Khalid Hussain Bhat, PhD (editor), Assistant professor, Division of Basic Sciences and Humanities, FOA, Wadura, Sopore, SKUAST-Kashmir.
Description: New York : Nova Science Publishers, [2022] | Series: Cell biology research progress | Includes bibliographical references and index. |
Identifiers: LCCN 2022040352 (print) | LCCN 2022040353 (ebook) | ISBN 9798886970449 (hardcover) | ISBN 9798886972030 (adobe pdf)
Subjects: LCSH: Genetic transcription. | RNA editing.
Classification: LCC QH450.2 .T6982 2022 (print) | LCC QH450.2 (ebook) | DDC 572.8/845--dc23/eng/20220831
LC record available at https://lccn.loc.gov/2022040352
LC ebook record available at https://lccn.loc.gov/2022040353

Published by Nova Science Publishers, Inc. † New York

Contents

Preface

The monumental finding of the “central dogma of molecular biology” by Watson and Crick made the world realize that at the molecular level the identity of a cell is specified by gene interactions. The transcription of a specific set of genes into RNA molecules at a given time (collectively known as “transcriptome”) specifies the identity of a cell and the final expression of this genetic information in the form of proteins regulates the biological activities within the cell. Living cells have been progressively refining and extending their molecular machinery and recording the results of their experiments in the genetic instructions they pass on to their progeny. Proper understanding of the genome therefore becomes essential for interpreting the functional elements of human biology and diseases. Central to the central dogma is the process of transcription, where the cell transcribes its RNA from the genome. The process of transcription in itself holds a lot of importance as it is the way by which hard-wired information in the DNA/genome can be decoded. Of the sea of RNAs transcribed by the cell, some have a function of their own, while others are translated to synthesize proteins.

This book is aimed to marvel at the information that transcriptome provides and how the development in the technology has transformed the vision of cell biology. We are amazed at the sophistication of mechanism and information we can gather from the enhanced technology. From the days when the identification of a single transcript was hailed to its progression where simultaneous existence of multiple transcripts in a population that varies with tissues and to the study of transcripts in a single cell is an indication of a great step forward. This book comprises three parts. Part I serves as an introduction to describe the development of transcriptome as a technique and the way by which it has enhanced our knowledge and have provided a better understanding of the basic biology of the cellular function. Part II of the book was written in an effort to understand the milestones and contributions of transcriptomics in understanding cancer. Lastly, Part III provides an insight on how transcriptomics is helping us to understand the microbes and host-pathogen interaction including bioremediation. Our goal for this book is that readers achieve a holistic view on how to put facts to use – to reason, predict, and control the behavior of living systems.

Chapter 1 - RNA (ribonucleic acid) molecules drive much of cellular actions. Messenger RNA (mRNA) relays genetic information from DNA to ribosomes to direct the formation of proteins. At the same time, noncoding RNAs act as enzymes, scaffolds, and regulatory molecules, contributing greatly to the regulation of gene activity, RNA splicing, protein synthesis, and even DNA replication. Transcriptomics studies the total RNA transcripts in the cell, encompassing both coding and non-coding RNA using high throughput sequencing techniques. Transcriptome analysis has become indispensable in basic and translational research. This chapter is aimed to set a baseline for the readers to understand the history of transcriptomics and its improvement with each passing technological advancement, the basics of the computational analysis and the potential of transcriptomics in shaping the future of science.

Chapter 2 - Despite the advancements made in medical science, cancer continues to present immense challenges worldwide as a dynamic, heterogeneous and lethal disease. With each passing day, awareness of its bio-molecular mechanisms, progresses and the path to cure is one of perseverance. Cancer is mechanistically driven by various genetic and epigenetic changes leading to the abnormal and disrupted expression of various critical genes. From time to time, extensive efforts have been made for the identification of these genetic alterations. However, various approaches in the form of cytogenetics, still-imaging, immuno-histological techniques and genomics for mutational screening have met with limited success. The limited knowledge hampers the proper understanding of these disease-causing mutations, culminating in the lack of agency in its detection, ultimately restricting proper clinical treatment. In response to these challenges, research efforts in the field of transcriptome biology have yielded a new hope in understanding various molecular abnormalities and alterations in different cancers.

Transcriptomics is a quantitative and qualitative branch of modern genetics that studies the total RNA transcripts in the cell, encompassing both coding and non-coding RNA using high throughput sequencing techniques. The qualitative gene expression profiling provides a more comprehensive understanding of genome, which allows a deeper insight into cancer mechanism by identifying molecular pathogenesis and numerous peculiar epigenetic irregularities involved in various malignancies. Transcriptomics has enabled the detection of even subtle changes in genetic profile expressions, which helps to determine differences in types/subtypes within the same cancer, assisting in its characterization, classification and molecular profiling. This data can be incorporated into different tumor molecular profiles, leading to identification of druggable targets, besides providing additional therapeutic options to the patients with advanced cancers. Moreover, molecular classification of cancer based on genomic and transcriptomic alterations have led to the identification of molecules (cancer biomarkers) that are specific to tumors. These biomarkers carry huge clinical impact, with critical insights in the cancer research for accurate and earlier detection of cancers, positively impacting diagnosis, prognosis and improved surgical and therapeutic interventions. Transcriptomics based gene expression tests for prediction of tumor behaviour and drug response are emerging as methods of choice.

In this chapter, the authors discuss the advent of transcriptomics and its impact in cancer research and molecular therapy. The authors highlight its advantages over conventional cytogenetic, histological and DNA based hybridization techniques. The authors bring to light the recent technological advancements in transcriptomics and discuss the potential of transcriptomics in shaping the future of precision oncology. Finally, the authors describe the apparent challenges and limitations of this technique like cost-benefit analysis, adequacy of mRNA abundance, fitness of mRNA etc for accurate interpretation of the transcriptomic data.

Chapter 3 - The main role played by DNA in the context of cells is the transferring of genetic information to daughter cells. While RNA can also act as an information carrier, its importance lies in the decoding of the information in the DNA and thus it is indispensable for normal function and growth of the cell. This property of RNA is fundamental to the process of the central dogma which states that information encoded in DNA is translated into proteins through mRNA. Although it seems straightforward, the variation in the stability and degradation rates of both RNA and proteins often result in differences in mRNA to protein expression in a time dependent and tissue specific manner. While the term transcriptome classically included mRNA, the template for protein synthesis, more than 95% of the cellular RNA is not translated into protein. However, these non-coding RNA transcripts have been well documented as key players in regulating the expression of many important signaling pathways and contribute towards maintenance of cellular homeostasis. Therefore, the term transcriptomics has since been amended to include all the RNA present in the cells including all forms of non-coding RNA.

Increasing evidence highlights the role of transcriptomics in identifying the function of genes from non-model organisms or the organisms from which the reference genome is still not available. Further, its role is inevitable in identifying the genes and signaling networks which undergo changes upon countering different environmental stress conditions. Transcriptome profiling provides us with valuable information such as identifying biomarkers for a disease, distinction between different stages or subtypes of disease and helps to identify the correlation between genetic profile and expression pattern in healthy vs diseased individuals. Transcriptomic profiling provides a valuable tool to identify fusion transcripts as well as novel transcripts especially in the context of carcinogenesis and thus could be beneficial in devising a personalized cancer therapeutic approach. Furthermore, transcriptome analysis can also provide a useful perspective on the functional genomics of neurological disorders as their etiology is otherwise very complex on account of their propensity to be influenced by various genetic and environmental factors.

Transcriptomics has a vast application in different aspects of cellular function and a closer examination of the applications will amplify our knowledge about the finely tuned genetic functions and how they modulate the signaling pathways and have biological implications. Over the last few decades, the advent of newer technologies has redefined the scope of investigation and its concurrent use with other omics approaches could provide us with a bigger understanding of complex cellular processes. Current knowledge on the role of transcriptomics in elucidating the cellular and molecular functions is summarized in this chapter.

Chapter 4 - Cancer cells are complex entities encompassing multiple cell types, resulting in molecular level heterogeneity leading to distinct and unequivocal clinical implications. Cancer stem cells (CSCs), a subset of specialized cells, play a critical role in the developing intratumoral complexities and functional diversity in these cells. The CSCs are described by self-renewal, multi-directional differentiation, inducing drug resistance, and carcinogenesis properties. Despite overall enhancement in cancer therapeutics, effectual therapy has been impeded by the manifestation of CSCs driven tumor cellular complexity that enterprise drug resistance and metastasis. Over the last decade, progress in medical sciences has steered towards the advancement of high throughput machineries such as single-cell transcriptomics (SCTs), which can analyse the gene expression and sequence report from individual cells of the minority population of cells including CSCs. SCTs provide the means to detect CSCs, study intratumoral heterogeneity, divulge the mechanism of drug resistance, and eventually direct personalized therapy approaches. Although high throughput technologies such as RNA sequencing provide a detailed understanding of cancer-related processes, these technologies are limited in that they represent "bulk" information from thousands of cancer cells. CSCs play critical roles and often control overall system behaviour, but they are not well reflected in "bulk" measurements, making them difficult to identify and quantify. Nonetheless, the absence of ubiquitously established CSC markers and characteristics has created disagreement among different reports. Identification and molecular characterization of CSCs, and the gene signalling pathways that regulate their behaviour and communications with other cell types, have substantial therapeutic potential to develop targeted therapies against cancer. Therefore, it becomes vital to use SCT technologies to comprehend the complexity of biological systems involved in cancer progression. In contrast to bulk transcriptomics, SCT technologies proffer a chance for the equitable identification and investigation of these CSC populations present in meagre quantities and a quiescent or latent condition and to develop specific cancer therapeutics. Finally, SCTs can identify minor but significant drug-resistant cancer stem cell populations in heterogeneous tumors, which can be utilized to select suitable treatments. However, SCTs have their competencies and inadequacies in terms of throughput, sensitivity, accuracy, and precision. In this chapter, the authors outline the role of SCTs in CSCs and offer their personal views about prospects that SCTs hold in the field. Singularizing tumor heterogeneity into noticeable morphological and phenotypic profiles continues to be a thought-provoking yet imperative goal in cancer treatment and management. Notably, the occurrence of SCTs in recent years has delivered an influential tool in resolving this impediment. Once employed in clinical samples amid tumor types, SCT technologies may prompt exponential growth and develop insights to explore the multitude of signalling pathways involved in carcinogenesis and contribute to developing strategies to combat drug resistance.

Chapter 5 - In the evolving technological world, the authors have made tremendous progress in biology and stepped past the 'omics era'. Omics refers to the collective analysis of large/meta data sets of biological variables comprising genes (genomics), RNAs (transcriptomics), protein (proteomics), lipids (lipidomics), epigenetics (epigenomics) and metabolites (metabolomics). One of the applications of omics includes the development of precision medicine which focuses on delivering personalized medication depending on the patient's need. Precision oncology aims at identifying the molecular markers of tumours to find out the therapeutics. Cancer is a disease of accumulating genetic and epigenetic changes in numerous genes, some of which drives tumour promotion and progression. These changes in the tumour microenvironment contribute to the phenotypic and functional heterogeneity within the same tumour making it resistant to the therapy. Transcriptomics comprises the expression of all the coding and non-coding RNA transcripts in an individual or population of cells providing a complete view of the expression dynamics. Transcriptomic studies give an insight into the gene expression variability in the tumours and their microenvironment. Different types of transcriptome technologies such as RNA sequencing, microarray and single-cell transcriptomics have evolved with time, and help us to design cancer precision medicine by identifying the actionable biomarkers. These techniques also act as good predictors of the response to the therapies providing an edge to both the oncologists and patients. Moreover, these techniques help to identify disease-specific biomarkers which facilitate diagnosis. These may also be used to classify cancer into clinically relevant molecular subtypes based on the variability in the expression profile of tumour cells. In future, increased knowledge in the field of precision medicine can lead to early diagnosis and better prognosis compared to other traditional approaches such as histopathology. Recent development in the field of cancer transcriptome has revealed extensive studies using RNA sequencing and generated databases for various cancer cell lines and tissue samples. These databases have greatly aided in the identification of combination of target-biomarkers, aiding to the development of precision medicine based on the different types and stages of cancer. Database like TCGA, CCLE, NCI60, GDSC, LINCS and Project Achilles have the compilation of data for multiple cancer types obtained from various cancer patients and cell lines through RNA sequencing and DNA microarray assay. These databases have the record of a number of mutations and expression profiles of various genes. Thus, the data from these databases can be used for the design of precision medicine based on the targetable mutations and gene expression patterns. Transcriptome engineering to manipulate the loss of function and gain of function changes in the gene expression at transcriptome level is of high significance in the development of specifically designed drugs for cancer treatment. This kind of engineering utilizes different transcriptome engineering tools such as siRNAs, shRNAs and CRISPR-Cas system. This chapter describes the recent developments in the field of cancer transcriptomics that can be translated into clinical oncology.

Chapter 6 - Host cells are exposed to a plethora of microbial pathogens such as viruses, bacteria, fungi, protozoans. Each microbial interaction with the host has a unique dynamics/interplay of events required for the survival, adaptation, and evolution of the pathogens as well as the host. Understanding these highly complex host-pathogen interactions are a requisite for a better understanding and prevention of infectious diseases. Continuous technological revolutions have refined our ability in flawless study of gene expression. The architecture of the gene expression network of these interactions could be decoded with high-throughput RNA sequencing. Transcriptomics technologies has become a powerful indispensable tool for the exploration of the association of a single-cell bacterium to the multicellular eukaryotic organism. Here, the authors describe the recent advancements and developments in transcriptomic approaches, involved in the exploration of host-pathogen interactions. Single-cell RNAseq and dual RNAseq are the most common and innovative approaches that have given us insights in the cell-to-cell interaction studies. Transcriptomic studies in infectious diseases also helped us identify the potential unique set of transcripts to be used as biomarkers for the disease. Moreover, the dual RNAseq has been successfully implemented in the study of various microbial pathogens that causes infection in humans and cereal plant cells such as Mycobacterium tuberculosis, Salmonella species, Xanthomonas oryzae, etc Further advancements in transcriptomic analysis require the integration of transcriptomics with the proteomics, metabolomics and genomics data along with the inclusion of other scientific disciplines like mathematical modelling. In conclusion, transcriptomics analysis of host-pathogen interactions reveals the dynamics of the molecular changes and helps in a better understanding of the disease and ways to combat its progression.

Chapter 7 - Effective restoration in environment is an important aspect in present day scenario and for that purpose the one among the best method considered is the bioremediation mediated by microbes, but inefficient knowledge regarding the same is a halt in the on field application of such techniques. Transcriptomics, proteomics, and interactomics, which are all relatively new approaches, hold a lot of potential as tools for addressing long-standing issues about the molecular mechanisms that drive mineralization routes. Transcript architecture and expression during mineralization have been examined utilising high-throughput transcriptomic approaches with microarrays. As far as the concern is related to the studying of mRNA at expression level, DNA microarray is considered to be one of the most sophisticated among others. Two-dimensional polyacrylamide gel electrophoresis in proteomics is a good method used in proteomics to determine the proteins (2-DE). For the construction of functional proteomics, the most recent breakthrough is the mass spectrometry and protein microarray. By far bioremediation mediated through use of differential expression can help in determining the whole profile of protein of microbes, essentially important for the purpose of bioremediation mediated by microbes.

Chapter 8 - Novel crops with higher production capacity, greater environmental resilience, and agro-biological characteristics that reduce chemical use are the need of the hour attributing to the market demand, climate change, and environmental concerns. Traditional breeding techniques can aid in the improvement of crop traits for instance that of improvement in quality, nutrition, as well as yield, but not at the necessary pace. They have been thoroughly studied to resolve the issue of curtailed yield in varied crops but haven't yielded any covet results. Next gen sequencing and through-put mechanism have approaches have unbolted vast research directions, also crop genomes have improved because of more profitable and advanced molecular breeding techniques. Sequencing-based methods, such as SAGE (Serial Study of Gene Expression), EST (Expressed sequence Tags Sequencing), MPSS (Massive Parallel Signature Sequencing), RNA-seq, and Hybridization-based approaches, such as Microarray Technology and Tiling-arrays, are some of the techniques widely used in plant breeding methods. Many transcriptomic methods such as DNA-Seq, RNA-Seq, ChIP-Seq, Methyl-Seq, MAINE-Seq, RIP-Seq and PhIT-Seq have advanced or vision about gene expression as well as gene regulation in plants. RNA-seq among these has emerged as the most efficient transcriptome tool. RNA-Seq also used to classify differentially expressed genes, has implications in functional genomics studies, particularly when the genomes being studied have limited knowledge. Crop development experiments focused on genomes and the obtainability of genomic data, as well as governing resources, tend to be unique to breeding programs. Many technological advances are being made which include expanded computing space, speedy processing, and improved investigative breakthrough and methods for downstream study and visualization. Because of the increased availability of genetic sequencing facilities and bioinformatics methods, analyses of individual genes are shifting to whole-genome research, which offers a more comprehensive view of how all genes interact. This could lead to a better sapience of gene functioning and potential regulatory issues within gene mesh, as well as more efficiently and quickly improved plants. The availability of plant genome sequences will assist with orphan crop research and use, as well as de novo domestication. In 21st century, "genomics-reinforced breeding" carries a lot of promise. In this chapter the authors discuss the transition from traditional breeding to modern genomic tools that will spur and quicken genomic-assisted breeding for the identification of avant-garde genes pertaining to increased yield and quality character resulting in long-term plant development.

Acknowledgments

We are indebted to many people towards the completion of this book. Full acknowledgment for scientific help is given separately. We owe a huge debt to Nova Publishers who helped convert the efforts of the authors into an informational asset. We thank them all for their help. We thank all the reviewers who peer reviewed the chapters and their suggestions added significantly to the book. Lastly, and most importantly, we thank our family for their constant support. Without their patience this book would not have come to a conclusion.

In writing this book we have been benefitted greatly by the inputs from many biologists. Their input has greatly enhanced the quality of the book. We would like to thank the following for their critical review and suggestions.

1) Nishant Singh, PhD
   Senior Scientist, Cell and Gene Therapy
   Absorption System, a Pharmaron company.
   Exton, PA-19335
   Email: nishant900@gmail.com

2) Nibedita Patel, PhD.,
   CRB1
   Jhons Hopkins Medicine
   1550 Orleans St
   Baltimore, MD 21287
   Email: npatel97@jhmi.edu

3) Sangeeta Kumari, PhD.,
   Pathology Advanced Translational Research Unit (PATRU)
   Pathology and Laboratory Medicine Department

Emory University, School of Medicine
Atlanta, GA 30322
Email: sangeeta.kumari@emory.edu

4) Sobia Zaidi, PhD.,
Department of Biomedical Sciences
Academic Research Center 378,
Heritage College of Osteopathic Medicine
Ohio University, Athens, Ohio
Email: zaidi@ohio.edu

5) Neeraj Soni, Ph.D.
Department of Cell and Developmental Biology
The University of Michigan, Ann Arbor, MI 48109
Email:sneeraj@med.umich.edu

6) Srijani Basu, PhD
Department of Medicine,
Weill Cornell Medicine,
New York, NY, USA.
Email: srb2003@med.cornell.edu

7) Mohammad Kamran, PhD
Department of Pathology and Laboratory Medicine
Weill Cornell Medicine,
New York NY 10021
Email: mok4005@med.cornell.edu

8) Anju Kumari, PhD
Thoracic Surgery branch, Centre for cancer research
National Cancer Institute, National Institutes of Health,
Bethesda, Maryland-USA
Email: anju.kumari@nih.gov

List of Acronyms

ABCG2	ATP-binding Cassette Super-family G member 2
ABI SOLID	Applied Biosystems, Small oligonucleotide ligation and detection system
ABL1	Abelson Murine Leukemia Viral Oncogene Homolog1
ADAR2	Adenosine Deaminase Acting on RNA 2
ADCY5	Adenylyl Cyclase Type V
AFF3	AF4/FMR2 Family Member 3
ALK	Anaplastic Lymphoma Kinase
AMHR2	Anti-Mullerian Hormone Receptor Type 2
AML	Acute Myeloid Leukemia
APOBEC	Apolipoprotein B mRNA Editing Catalytic Polypeptide-Like
ARID1A	AT-Rich Interactive Domain-Containing Protein 1A
ATLL	Adult T-cell Lymphoma
ATP1A1/2	ATPase Na+/K+ Transporting Subunit Alpha 1/2
BCL2	B-cell lymphoma 2
BGJ398	Infigratinib
BID	BH3 Interacting Domain Death Agonist
BMDM	Bone Marrow Derived Macrophages
BPH	Benign Hyperplasia
BRPF1	Bromodomain and PHD Finger-Containing Protein 1
CA 19-9	Carbohydrate Antigen 19-9
CAFs	Cancer-Associated Fibroblasts
CAGE	Cap Analysis Gene Expression
CALLA 01	Cyclodextrin-Based Nanoparticle System Containing SiRNA Against M2 Subunit of Ribonucleotide Reductase
CC	Cervical cancer
CC11orf86	Chromosome 11 Open Reading Frame 86

cCAFs	Cycling CAFs
CCC	Clear Cell Carcinoma
CCLE	The Broad Novartis Cancer Cell Line Encyclopedia
CCR5	C-C Motif Chemokine Receptor 5
CDKN2A	Cyclin-Dependent Kinase Inhibitor 2A
CE	Capillary electrophoresis
CEACAM1	Carcinoembryonic antigen-related cell adhesion molecule 1
CEBPβ	CCAAT enhancer binding protein
CEL-seq	Cell Expression by Linear Amplification and Sequencing
CHI3L1	Chitinase-3-like protein 1
CITE-seq	cellular indexing of transcriptomes and epitopes by sequencing
CL	Clinical subtype
CLL	Chronic Lymphocytic Leukemia
CML	Chronic Myeloid Leukemia
CMS	Consensus Molecular Subtypes
CMTM5	CKLF Like MARVEL Transmembrane Domain Containing 5
CNV	Copy Number Variations
CRC	Colorectal Cancer
cSCC	Cutaneous Squamous Cell Carcinoma
CSFR1	Colony-Stimulating Factor 1 Receptor
CTC	Circulating Tumor Cells
CTCL	Cutaneous T cell Lymphoma
CTLA4	Cytotoxic T-Lymphocyte Associated Protein 4
CWDEs	Cell Wall Degrading Enzymes
CXCL5/11	C-X-C Motif Chemokine Ligand 5/11
CYP2A7	Cytochrome P450 Family 2 Subfamily A Member 7
DCIS	Ductal Carcinoma in situ
DCs	dendritic cells
DEG	Differentially Expressed Genes
DIMS	Direct injection mass spectrometry
DLBCL	Diffuse large B-cell Lymphoma
DOPC	Dioleoyl phosphatidylcholine
DroNC-seq	Droplet Single-Nucleus RNA Sequencing
DTC	Differentiated Thyroid Cancer
EC	Endothelial cells

EGFR	Epidermal Growth Factor Receptor
EML2	Echinoderm Microtubule-Associated Protein-Like 2
EMT	Epithelial To Mesenchymal Transition
ENL	Eleven-Nineteen Leukemia
EOC	Epithelial Ovarian Cancer
EpCAM	Epithelial Cellular Adhesion Molecule
EphA2	Ephrin Type-A Receptor 2
eQTL	Expression quantitative trait loci
ER	Estrogen Receptor
ERBB2	erbB-2 (Erythroblastic Oncogene B) Receptor tyrosine-protein kinase
ESCC	Esophageal Squamous Cell Carcinoma
ESTs	Expressed Sequence Tags
ETV4	ETS Translocation Variant 4
FACS	Fluorescence Activated Cell Sorting
FAK	Focal Adhesion Kinase
FAO	Food & Agricultural Organization
FDA	Food and Drug Administration
FGFRL1	Fibroblast Growth Factor Receptor-Like 1
FISSEQ	fluorescent in situ sequencing
FN1	Fibronectin 1
FNA	Fine-needle aspiration
FOXM1	Forkhead Box M1
FSD1	Fibronectin Type III and SPRY Domain Containing 1
FSIP1	Fibrous Sheath-Interacting Protein 1
FT-IR	Fourier transform-infrared spectroscopy
FTH1	Ferritin Heavy Chain 1
GATA4	GATA Binding Protein 4
GDSC	The Genomics of Drug Sensitivity in Cancer
GOF	Gain of function
GOI	Genes of Interest
GTEx	Genotype Tissue Expression
GWAS	Genome-wide Association Studies
HAND	HIV-associated neurocognitive disorder
HCC	Hepatocellular Carcinoma
HE4	Human Epididymal Secretory Protein E4
HER2	Human Epidermal Growth Factor Receptor 2
HGSOC	High Grade Serous Ovarian Cancer

HMSD	Histocompatibility Minor Serpin Domain
HNnov-miRNAs	Head and Neck Specific Novel microRNA
HNSCC	Head and Neck Squamous Cell Carcinoma
HPRRs	Host Pattern Recognition Receptors
HPV	Human Papillomavirus
HSC	Hematopoietic Stem Cell
ICE	Inosine Chemical Erasing
IDO1	Indoleamine 2,3-Dioxygenase 1
IER3	Immediate early response 3
IFIT3	Interferon Induced Protein with Tetratricopeptide Repeats 3
IICs	Infiltrating immune cells
IL21R	Interleukin 21 Receptor
IRGs	Immune-Related GenesOS-Overall Survival
ISGI5	Interferon-Stimulated 15kd Gene
JA	Jasmonic Acid
JAK2	Janus Kinase 2
KAT2A	Lysine Acetyltransferase 2A
KRAB	Krüppel-Associated Box
KRAS	Ki-Ras2 Kirsten Rat Sarcoma Viral Oncogene Homolog
LCM	Laser Capture Microdissection
LGSOC	Low Grade Ovarian Serous Carcinoma
LINCS	The Library of Integrated Network-Based Cellular Signatures
LNA	Locked Nucleic Acid
lncRNAs	Long non-coding RNAs
LOF	Loss of function
LOX	Lysyl Oxidase
LSC	Leukemia Stem Cell
LSCC	Laryngeal Squamous Cell Carcinoma
LSR	Lipolysis Stimulated Lipoprotein Receptor
LY75	Lymphocyte Antigen 75
MAPK	Mitogen-Activated Protein Kinase
MAPK/ERK	Mitogen-Activated Protein Kinase /Extracellular Signal-Regulated Kinase 1/2
MARS-seq	Massively parallel RNA Single-cell Sequencing

MATQ-seq	Multiple Annealing and Tailing-based Quantitative single cell RNA sequencing
mCAFs	Matrix CAFs
MCL1	Myeloid Cell Leukemia 1
MERFISH	multiplexed error-robust fluorescence in situ hybridization
MET	Mesenchymal-to-Epithelial Transition
microSPLiT	microbial split-pool ligation transcriptomics
mIHC	multiplex immunohistochemistry
miR	microRNA
MLL-AF4	Mixed Lineage Leukemia (MLL) Gene Harboring Acute Lymphoblastic Leukemia (ALL1)-Fused Gene from Chromosome 4 (AF4) Fusion Proteins
MLLT1	Mixed Lineage Leukemia Translocated to 1
MMP3	Matrix Metallopeptidase 3
MMSEQ	multi-mapping RNA-sequencing
MMTV-PyMT	Mouse Mammary Tumor Virus-Polyoma Middle Tumor-antigen
MPSS	Massive Parallel Signature Sequencing
mRNA	messenger RNA
MS	Mesenchymal Subtype
mTOR	Mammalian Target Of Rapamycin
NAMs	network-attacking mutations
NCBI	National Centre for Biotechnology Informatics
NCI	National Cancer Institute
ncRNA	Noncoding RNA
NCT	National Clinical Trials
NECs	Normal Endothelial cells
NFi	Nuclear Factor I
NGS	Next-Generation Sequencing
NK	Natural Killer
NMR	Nuclear magnetic resonance
NPM1	Nucleophosmin Gene
NSCLC	Non-Small Cell Lung Carcinoma
NTRK1/3	Neurotrophic Tyrosine Kinase Receptor Type I/III
OCSCC	Oral Cavity Squamous Cell Carcinoma
OSBP	Oxysterol-Binding Protein 1
PAK1	P21 (RAC1) Activated Kinase 1
PAMAM	Poly(amidoamine)

PAMPs	Pathogen Associated Molecular Patterns
PAX5	Paired Box 5
PCa	Prostate cancer
PCAWG	Pan-Cancer Analysis of Whole Genomes
pcRNA	Protein-Coding Ribonucleic Acid
PCSK6	Proprotein Convertase Subtilisin/Kexin Type 6
PD-L1	Programmed Death-Ligand 1
PEI	Polyethyleneimine
PETRI-seq	Prokaryotic Expression profiling by Tagging RNA In situ and sequencing
PHC3	Polyhomeotic-like Protein 3
PiRNA	Piwi interacting RNA
Plk1	Polo Like Kinase 1
PR	Progesterone Receptor
PRKACA	Protein Kinase Camp-Activated Catalytic Subunit Alpha
PSA	Prostate specific antigen
PSAT1	Phosphoserine Aminotransferase 1
PSMA	Prostate-Specific Membrane Antigen
PTEN	Phosphatase and Tensin Homolog
PTPN11(Shp2)	Protein Tyrosine Phosphatase Non-Receptor Type 11
PVR	Polio Virus Receptor
qRT-PCR	Real-Time Quantitative Reverse Transcription PCR
QTLs	Quantitative Trait Lockers
RAB11FIP4	RAB11 Family Interacting Protein 4
Raf/MEK/ERK	Rapidly Accelerated Fibrosarcoma (Raf) /Mitogen-activated protein kinase (MEK)/ Extracellular-Signal-Regulated Kinase (ERK)
RAGE	Repertoire and Gene Expression
RBM10	RNA Binding Motif Protein 10
RCC	Renal Cell Carcinoma
REAP-seq	RNA expression and protein sequencing assay
REPAIR	RNA Editing for Programmable Adenosine to Inosine Replacement
RIP-Seq	RNA immunoprecipitation sequence
RIPK2	Receptor-Interacting Serine/Threonine-Protein Kinase 2
RNAi	RNA Interference

rRNA	ribosomal RNA
RSAD2	Radical S-Adenosyl Methionine Domain Containing 2
SA	Salicylic Acid
SAGE	Serial Analysis Gene Analysis
SCC-Ag	Serum squamous cell carcinoma antigen
SCCHN	Squamous Cell Carcinoma Of The Head And Neck
sci-CAR	single-cell combinatorial indexing chromatin accessibility
scM&T-seq	single-cell methylome and transcriptome sequencing
scRNA-seq	Single-cell RNA sequencing
SCT	Single Cell Transcriptomics
SEPT9	Septin 9
SFRP2	Secreted Frizzled Related Protein 2
sgRNA	Single Guide RNA
SHERLOCK	Specific High Sensitivity Enzymatic Reporter Unlocking
shRNAs	Short hairpin RNA
siRNAs	Small interfering RNA
SL-seq	Spliced Leader RNA-seq
Smart-seq	Switch Mechanism at the 5′ End of RNA Templates Sequencing
SMD	Sterility Mosaic Disease
snoRNAs	Small nucleolar RNAs
SNP	Single nucleotide polymorphism
snRNAs	small nuclear ribonucleoproteins
SOCS1	Suppressor of Cytokine Signaling 1
SOS	Son of Sevenless
SPLiT-seq	Split-Pool Ligation-based Transcriptome Sequencing
SPP1	Secreted Phosphoprotein 1
STAT1	Signal Transducer and Activator of Transcription 1
STK1/11	Serine/Threonine Kinase 1/11
SUM1315	Triple negative breast cancer (TNBC) cell line
TCGA	The Cancer Genome Atlas
TdECs	Tumor-Derived Endothelial Cells
TECs	Tumor Endothelial cells
TFF1	Trefoil Factor 1
TFRC	Transferrin Receptor

TI	Transcription interference
TLK2	Tousled Like Kinase
TMB	Tumor Mutation Burden
TME	Tumor Microenvironment
TNBC	Triple Negative Breast Cancer
TNFRSF12A	TNF Receptor Superfamily Member 12A
TNM	Tumor Node Metastasis
TP63	Tumor Protein P63
TROY	Tumor necrosis factor receptor superfamily member 19
TS miRNA	Tumor Suppressor microRNAs
TSK	tumor-specific keratinocyte
TWIST1	Twist Family Basic Helix Loop Helix Transcription Factor 1
UACA	Uveal Autoantigen with Coiled-Coil Domains, Ankyrin Repeats
UMBBD	University of Minnesota Bio catalysis /Biodegradation Database
VEGF-A	Vascular endothelial growth factor A
VPR	VP64-p65-Rta Activator
WINTHER	Worldwide Innovative Networking Therapeutics
Wnt	Wingless Related Integration Site

Part I. Introduction to Transcriptomics

Chapter 1

The Nuts and Bolts in Transcriptomics

Sarah Naiyer[1,*] and Khalid Hussain Bhat[2]
[1]Department of Microbiology and Immunology,
University of Illinois at Chicago, Il, USA
[2]Division of Basic Sciences and Humanities, FOA, Wadura, Sopore,
SKUAST-Kashmir, India

Abstract

RNA (ribonucleic acid) molecules drive much of cellular actions. Messenger RNA (mRNA) relays genetic information from DNA to ribosomes to direct the formation of proteins. At the same time, noncoding RNAs act as enzymes, scaffolds, and regulatory molecules, contributing greatly to the regulation of gene activity, RNA splicing, protein synthesis, and even DNA replication. Transcriptomics studies the total RNA transcripts in the cell, encompassing both coding and non-coding RNA using high throughput sequencing techniques. Transcriptome analysis has become indispensable in basic and translational research. This chapter is aimed to set a baseline for the readers to understand the history of transcriptomics and its improvement with each passing technological advancement, the basics of the computational analysis and the potential of transcriptomics in shaping the future of science.

Keywords: RNA, transcriptomics, central dogma, translational research

---

*Corresponding Author's Email: sara.naiyer@gmail.com.

In: Transcriptomics and their Importance
Editors: Sarah Naiyer and. Khalid Hussain Bhat
ISBN: 979-8-88697-044-9

1. Introduction

Central to the central dogma is the process of transcription, where the cell transcribes its RNA from the genome. The process of transcription holds a lot of importance as it is the only method till date by which the genome can be decoded. To define, transcriptome is the sum of all the RNA transcripts of a cell, transcriptomics is the study of transcriptome and its functions, and the techniques used to study an organism's transcriptome are called the transcriptomic technologies. Transcriptome profiling identifies genetic expression pattern of a cell that helps in understanding the genetic regulation of a particular cell type. Of the sea of RNAs transcribed by the cell only a fraction is translated and diverted to protein synthesis while most have a function of their own. RNA represents the first quantitative link between the genome sequence and the phenotype of the concerned organism/species. Some of the very popular traditional ways to study an individual transcript or a group of transcripts include low throughput sanger sequencing, northern blot and real time PCR (Alwine, Kemp, and Stark 1977). This allows the study of the expression levels of a concerned transcript. However, these methods are very tedious and time consuming and provide limited results. A lot of information concerning the interplay between different transcripts cannot be deciphered by mere traditional approaches. As the different pathways of a cell depend intriguingly on the cross signaling among different transcripts, studying limited transcripts would never give a complete picture of the interrelated and inter dependent functions of the individuals. The urge to understand the role of different transcripts to unravel cell biology introduced ways to study multiple transcripts together. In this chapter we aim to discuss the history of transcriptomics, the current computational ways for data analysis and the future methodology and aspects of transcriptome analysis.

2. History towards the Development of Transcriptomics

To study multiple transcripts together at the same time the field of transcriptomics emerged at around 1990s. As a step towards transcriptomic studies, during late 1970s cDNA libraries were generated using reverse transcriptase followed by Sanger sequencing during the next decade (Sutcliffe et al. 1982; Pan et al. 2008; MA, L, and RH 1998). The first attempt to study the human transcriptome was done in 1991, where the authors reported a study of 609 mRNA transcripts from the human brain using Expressed sequence tags (ESTs) (Adams MD, Kelley JM, Gocayne JD, Dubnick M, Polymeropoulos MH, Xiao H, Merril CR, Wu A, Olde B 1991). ESTs are small DNA sequence (usually 200 to 400 nucleotides long) generated from a single RNA. In late 1970s, libraries of silk moth mRNAs were collected and converted into complementary DNA (Sim et al. 1979). The cDNA is used to generate EST by sequencing either one or both ends of the expressed gene (Parkinson and Blaxter 2009). ESTs became prominent during 1990s where in the 3' and 5' ends of transcripts were sequenced from the cDNA library to determine the expression level of the transcripts. Since then, there has been an exponential advancement in the development of methods suitable for transcriptome analysis. EST is a powerful tool in the process of gene discovery, and in addition, they provide an alternative to full-length cDNA sequencing (Ali et al. 2020). EST databases are used in the field of phylogenetics furnishing information regarding the evolutionary relationship between organisms. Furthermore, in proteomics they can also provide regions coding for potentially new proteins (Wu et al. 2002).

In 1995, the first sequence based transcriptome method called Serial Analysis of Gene Expression (SAGE) was developed in which the transcripts were restriction digested to obtain an 11-15nt transcript each of which were further concatenated to form a long transcript sequenced by sanger sequencing (Velculescu et al. 2009). One of the shortcomings of SAGE include its short tag size which poses a problem for organism of complex genome. SAGE tag primers are too short in determining the appropriate amplification conditions of cDNA from each gene. Also, the 15 bp tag was not suited as a probe for cDNA library screening. Different variants of SAGE "SuperSAGE" and LongSAGE were followed to improve the tag size (Matsumura et al. 2003, 2006) however, the method is time consuming and laborious (Lowe et al. 2017).

Another variant Cage Analysis of Gene Expression (CAGE) was also used (Shiraki et al. 2003), and is a very sensitive and accurate method for quantitative transcriptome analysis (TSS). CAGE, unlike Microarray and RNA-seq, can correctly identify transcriptional start sites (TSSs) and their associated promoter regions by sequencing the 3' end of cDNA (5' end of RNA). Thus, CAGE became a useful tool for studying gene regulation at the TSS level, allowing researchers to investigate genes controlled by numerous alternative promoters (Shiraki et al. 2003; Kodzius et al. 2006). Massively parallel signature sequencing (MPSS) was used in Arabidopsis for whole genome transcriptional analysis in 2004 (Meyers et al. 2004). However, the sequence tags generated in SAGE/CAGE/MPSS fall short of being uniquely mapped to the reference genome and only a portion of transcript is analyzed. Consequent upon these shortfalls, these traditional sequencing technologies do not provide a comprehensive transcriptome profile and gene annotation. In mid 1990s and 2000s, some of the prominent and popular contemporary techniques microarray and RNA Seq were developed (Nelson 2001; Wang, Gerstein, and Snyder 2009). DNA microarray is a nucleic acid hybridization technique that allows specific DNA sequences to be either deposited or synthesized on a surface (immobilization) in such a way that the DNA is covalently or noncovalently attached to the surface, referred to as 'DNA array.' The DNA array is typically used to probe a mixture of labelled nucleic acids by their hybridization to the 'probes' immobilized on that array. It is a cost effective and labor-saving gene discovery method (Heller 2002). With further advances in array manufacturing and designing, more and more genes were tested on a single array. DNA microarray has undergone significant progress as far as its technology and application is concerned since its advent in 1995 (Schena et al. 1995). However, some of the shortcomings of microarray included, relying on previous knowledge of genomes, cross hybridization causing high background levels and the narrow range of detection due to signal saturation and background. The advent of fluorescence detection increased the sensitivity and accuracy for low abundance transcripts (Pozhitkov, Tautz, and Noble 2007). But since, microarray required the probe to hybridize to the transcripts it could not provide information about any novel gene. Different variants of microarray spotted oligonucleotide array and Affymetrix high density arrays are used. However, with the emergence of next-generation high throughput sequencing technologies rapid and enormous advances in the transcriptomics field have been made, where RNA Seq has supplanting microarrays as the method of choice for transcriptional profiling. 454 technology published in 2006 was one of the earliest RNA seq works. Later in 2008 RNA seq became very popular, with the Solexa/Illumina technology which allowed the assay of 10^9 transcripts at the same time. It was Next Generation Sequencing (NGS) because of which RNA Seq boomed as it allowed the read out of a billion of transcripts at the same time. NGS is a massively parallel sequencing technology, capable of sequencing millions of fragments simultaneously in a single run with very high sequence coverage and resolution. NGS is equipped with massive advancement in detection systems and automation which has greatly increased its speed and has very effectively reduced the cost, and thus overcame the major limitation of First-generation sequencing (Grada and Weinbrecht 2013). It can also quantitatively and qualitatively assay any RNA type including mRNAs, miRNAs, siRNAs and lncRNAs.

3. RNA Sequencing (RNA-Seq)

RNA-Seq is an advanced sequencing approach for analysis and measurement of transcriptome of a given sample using high throughput next generation sequencing. The introduction of RNA-seq has revolutionized transcriptomic analysis and has enabled more comprehensive and detailed analysis of gene expression which were not possible to be detected by conventional sequencing approaches. RNA-Seq gives significantly more coverage and resolution of the transcriptome than earlier Sanger sequencing and microarray-based approaches. Beyond quantifying gene expression, the data generated by RNA-Seq facilitate the discovery of novel transcripts, non-protein-coding RNAs, identification of alternatively spliced genes, detection of allele specific expression and sites of RNA and protein interactions (Chatterjee et al. 2018; Dündar, Skrabanek, and Zumbo 2018). Some of the applications of RNA Seq are citied here (Sarah Naiyer et al. 2019; S. Naiyer, Bhattacharya, and Bhattacharya 2019; C. Li et al. 2021; Sarah Naiyer et al. 2022).

3.1. Advances in RNA-Seq

Since its inception more than a decade ago, there has been continuous evolution in the RNA sequencing methodologies, which has guided us towards a better understanding of crucial aspects of transcriptome biology, some of which are briefly outlined below and depicted in Figure 1:

1. Long-read cDNA sequencing: Long read cDNA sequencing technology developed by Pacific Biosciences (PacBio) and Oxford Nanopore (ONT) enables single-molecule sequencing of complete individual RNA molecules.
2. RNA sequencing and TSS mapping: The mapping of transcription start sites (TSSs) at nucleotide resolution define the RNA product fully and help identify adjacent promoter regions that regulate the expression of each transcript. CAGE was the first method adopted for the identification of TSSs. However, high RNA input requirements

and generation of short cDNA tags (lesser than 20 nucleotides per TSS) restricted its use. Combination of CAGE with deep sequencing RNA techniques including deepCAGe10, PeAT11, nanoCAGe and CAGescan have fairly contributed in overcoming such limitations. The development of HeliScopeCAGE protocol (Kanamori-Katayama et al. 2011) abolished the requirements like second strand synthesis, adapter ligation, digestion and PCR further simplifying the TSSs mapping procedures.

3. Characterization of alternate splicing: EST was the most common way of analysing spliced variants. Since it was costly and less efficient, with the improvement in RNA sequencing-based approaches it is now possible to get a more comprehensive and deeper overview of transcriptome thus allowing to detect even rare alternative splice variants. The combination of RNA-seq with improved read mapping algorithms such as TopHat (Langmead et al. 2009), RASL-Seq and improvements in "long read RNA sequencing" has greatly simplified the approaches towards the study of novel transcripts (Hairi Li, Qiu, and Fu 2012; Au et al. 2010).
4. Detection of gene fusion events: RNA-seq is preferably used for detecting events producing aberrant RNA species, more likely associated with various biological disorders. Gene fusion events can be detected, by combining RNA-Seq data with specific bioinformatic approaches, and by modifications in conventional RNA-Sequencing techniques. Hybrid sequencing, combining, SMRT (Single-molecule real-time sequencing) with short RNA-seq reads, has been used to detect fusion transcripts. SLR-RNA-Seq (synthetic long read RNA sequencing) has been used to detect both alternate splicing and aberrant gene fusion transcripts (Tilgner et al. 2015). The development of targeted RNA-seq techniques, like RNA CaptureSeq has further eased our way towards the detection of novel fusion transcripts (Mercer et al. 2014).
5. Targeted RNA Sequencing: This approach becomes method of choice for detecting transcripts with a very low expression profile and, for detecting, set of genes of particular interest. Targeted sequencing is based on two general approaches, namely: "*target capture*" in which specific RNA is targeted by using biotinylated probes and "*amplicon sequencing,*" which amplifies target cDNA using, gene specific primers (Hrdlickova, Toloue, and Tian 2017). Recently a target RNA sequencing approach known as TORNADO-seq has been used for

drug discovery in organoids, which are specific for colorectal cancers (Norkin, Ordóñez-Morán, and Huelsken 2021).

6. Profiling Small non-coding RNAs (sncRNAs): Advancement in NGSs, bioinformatics and computational analysis, has immensely contributed to the rapid progress in profiling of small non coding RNAs per se miRNAs, siRNAs, piRNAs, tRNA derived fragments, snoRNAs, vault RNA, rRNAs and Y RNAs. Recently, Kugelberg et al, demonstrated an effective and simple sequencing procedure (5′XPSRNA-seq) to sequence small RNAs (Kugelberg et al. 2021). A novel class of short PIWI-interacting RNAs has been revealed in human oocytes by using Single-cell CAS-seq (Yang et al. 2019).
7. Single cell RNA Seq: scRNA-seq aims to sequence and examine the transcriptomic information of thousands of individual cells in a single experiment, therefore providing a remarkable opportunity to explore the transcriptomic profiling of cells at a single cell resolution. Single Cell RNA sequencing is proving to be a powerful technique for revealing cellular heterogeneity and discovering novel cell types, thus carrying a huge potential in both basic biological and clinical research. The first scRNA-seq technique, was used in a four-cell stage blastomere (F. Tang et al. 2009), and since then, continuous improvements, in scRNA-seq, combined with advancements in molecular biology, microfluids and nanotechnology (Papalexi and Satija 2018) have resulted in the development of high resolution, single cell, transcriptomic sequencing approaches, vastly enhancing our ability to profile transcriptomics of single cells. Some of the following chapters have discussed scRNA in detail.
8. Dual RNA Seq: The combined total RNA isolated from the host and its interacting microbial partner is simultaneously sequenced and analysed. Dual RNA-seq was initially applied for the viral, fungal and parasite infection and shows the transcriptome architecture similarity with its host. The utmost advantage of dual-RNA seq is the high throughput expression analysis of transcripts that expressed during the interaction of host and microbial pathogen (Nuss et al. 2017).
9. Direct RNA Sequencing: Direct RNA synthesis technology, allows the sequencing of RNA directly, with extremely low input RNA and doesn't require the conversion of RNA to complementary DNA, and other ligation and amplification steps required for RNA sequencing, hence overcoming the limitations posed by cDNA synthesis. However, this direct sequencing technique generates short sequence

reads which limits its use in analysis of some aspects of transcriptomics e.g, proper identification the alternate splicing in eukaryotes. The recent development of parallel direct RNA sequencing (nanopore direct RNA sequencing) enables the sequencing and analysis of full-length transcripts (Garalde et al. 2018).

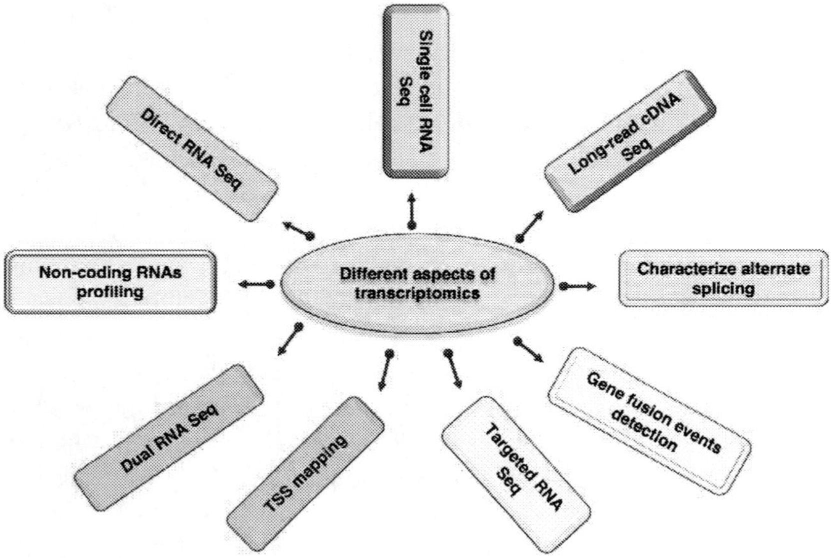

Figure 1. Different sub-branches of transcriptomics.

4. Computational Analysis of RNA Sequencing Data

The number of computational research methods for RNA sequencing has grown significantly over the last decade. The selection of a method depends on the objective and precision of the experiment (Soverini et al. 2019; Chatterjee et al. 2018). The four basic steps of a generalized RNA sequencing data analysis procedure are shown in Figure 2 and involve:

4.1. Quality Control of Raw Data

The raw sequencing data, which is normally given in the form of FASTQ files (Cock et al. 2010), is assessed and cleaned as the first step in data analysis. Quality management visibly illustrates the sequencing quality by excluding low-quality readings, removing low-quality bases, and trimming adaptor sequences (Conesa et al. 2016).

Some of the popular resources used for this purpose include RNA-SeQC (Bolger et al. 2018), FASTQ (S. Chen et al. 2018), PRINSEQ (Schmieder and Edwards 2011), SOAPnuke (Y. Chen et al. 2018), NGSQC (Patel and Jain 2012), and Trimmomatic (Bolger, Lohse, and Usadel 2014).

4.2. Read Alignment and Transcript Assembly

The clean reads must then be mapped to either a transcriptome or a genome. Bowtie (Langmead et al. 2009), STAR (Dobin et al. 2013), Tophat2 (Kim et al. 2013), BWA (H. Li and Durbin 2009), and HISAT2 (Kim, Langmead, and Salzberg 2015) are some of the mapping resources available. This is followed by compilation of transcripts from short reads. SOAPdenovoTrans (Hurgobin 2016), Trans-AByS (Robertson et al. 2010), StringTie (Pertea et al. 2015), Cufflinks (Trapnell et al. 2010), and Trinity (Grabherr et al. 2011), and are some of the examples of applications that can be used for assembling transcripts.

4.3. Expression Quantification

The establishing of the transcript model allows its expression to be monitored at genome, exon or transcript level. Genome level quantification software include HTSeq-count (Anders, Pyl, and Huber 2015), and FeatureCount (Liao, Smyth, and Shi 2014). Exon level quantification software is DEXSeq (Anders, Reyes, and Huber 2012) whereas, eXpress (Roberts and Pachter 2013), RSEM (B. Li and Dewey 2011) and Cufflinks (Trapnell et al. 2010) are some of the commonly used transcript level quantitative software.

4.4. Differential Expression Analysis

After normalization, an expression matrix is developed, which can be used to recognize differentially expressed genes using statistical methods. This task can be performed by using different platforms including EdgeR (Robinson, McCarthy, and Smyth 2010), DESeq2 (Love, Huber, and Anders 2014) EBSeq (Leng et al. 2013), etc.

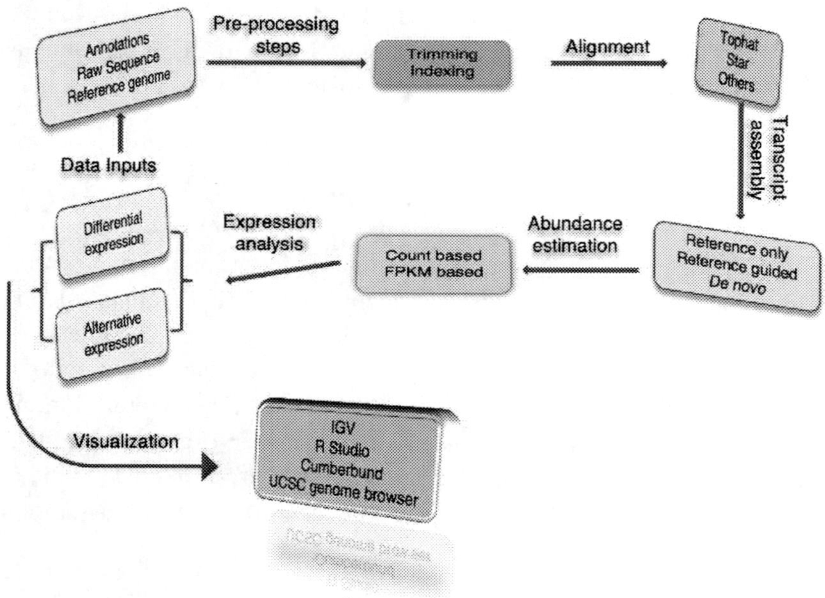

Figure 2. A schematic depicting the steps involved in RNA-seq analysis.

Conclusion and Further Advancements in Transcriptomic Analysis

Despite the marked improvements in RNA sequencing methodologies, there is still plenty of room left for further innovations regarding profiling and characterization of the transcriptome. Cellular RNA analysis holds lots of promises for the future which includes its use in clinical applications like identifying drug targets, disease biomarkers and identifying good or poor respondents to a drug. The Oxford Nanopore Technologies developed a sequencing device that uses nanopore technology, abbv. as MinION, which is further a big advancement in the field. This device consists of a nanopore based device where pores are inserted in a membrane kept over an electrical detection grid. When the DNA molecules pass through the pores, it generates a current. A base calling algorithm detects the variations in the sequences and infers the sequence of the nucleotides. Under perfect conditions, first the leading strand makes its way through the pore which is followed by the hairpin adapter and lastly, the complement strand. In the scenario where both strands are sequenced, a consensus sequence called "2D reads" is generated. These reads have high accuracy compared to reads that involve a single pass of the molecule called "1D reads." Long read lengths generated from a handheld sequencer can be used in many important applanations including but not limited to genomics, *de novo* genome assembly of new genomes, analysis of variations in healthy versus diseased samples and resolution of isoforms in cDNA sequencing. Goodwin et al. developed a novel algorithm called NANOCORR given that the existing packages most often than not can't put together long reads and have high error rates. They conclude that the long reads obtained using the nanopore sequencer give a holistic view/representation of the genome and also helps with the correct assemblage of the gene cassettes, transposable elements, and rRNAs (Goodwin et al. 2015).

Another technology called bead Array technology by illumine uses silica microbead giving way to sequencing of millions of genotypes for a single individual at once. Cell free gene expression (*in vitro*) is also used now a days and has application in synthetic biology, biotechnology, and biomedicine. Scientist have used *E. coli* based cell free expression system to regulate gene expression in time and space (Finkler and Ott 2019). Developmental and disease mechanisms are most often cell type–specific processes. Single-cell RNAseq (scRNA-seq) method allows simultaneous expression profiling of the transcriptome of thousands of individual cells, overcoming many issues linked to tissue specificity and bulk RNA Sequencing thus revolutionizing our understanding of cell type-specific processes *in vivo*. Subsequent improvements in biochemical methods in conjunction with the breakthroughs in next-generation sequencing have led to the current revolution in scRNA-seq. The application of RNA seq method to single-cell transcriptional profiling has enabled more precise characterization of cell lineage and cell state genetic profiles. The development of bioinformatic and statistical tools has provided methods for differential gene expression analysis, RNA isoform analysis, haplotype-specific analysis of gene expression (allele-specific expression), and analysis of expression quantitative trait loci. Advancements in scRNA-seq to generate less waste and make it comparatively easy are a constant process. Magnetic bead-assisted parallel single-cell gene expression sequencing (MAPS-seq) is a method that increases the ease of sample preparation and reduces reagent waste (Park et al. 2020). 237 human myelogenous leukemia cells treated with different drugs were analyzed using this method. Development in microfluidics to separate cells for single cell transcriptome analysis (Yin and Marshall 2012) has been incredibly helpful and has been discussed in detail in later chapters.

A recent technology called spatial transcriptomics teases apart the gene expression in context of spatial location of each cell trying to answer the question of why some diseased/cancer cells respond to imuunotherapy and others do not. However, the sensitivity of very low copy number mRNA is not yet established, the coverage is still not extended to miRNA and lncRNA and the normalization methods used are variable. This advancent holds great promises for personalized medicines which is a topic of discussion in later chapters.

Advancements in other aspects of total cellular RNA pool, that include the total circular RNA is still lagging and holds a lot of potential in unrevealing multiple disease mechanisms. These are large class of non-coding RNA synthesized during splicing and are potential biomarkers for some diseases including Alzheimer's and certain cancers till date (Tang et al. 2020; Kristensen et al. 2019). Scientists are struggling to identify total cellular circular RNA in the cell and may further want to classify them in space and time. It holds lots of promises to explore the roles of these non-coding RNAs via transcriptomics in regulating cellular dynamics in diseased conditions. The explosion and ever-growing advancement in the field of RNA biology and transcriptomics holds enormous clinical utility in profiling human disease. We hope the readers of the book explore different aspects of transcriptome biology and widen their horizon of the enoromous potential it holds.

References

Adams MD, Kelley JM, Gocayne JD, Dubnick M, Polymeropoulos MH, Xiao H, Merril CR, Wu A, Olde B, Moreno RF. 1991. "Complementary DNA Sequencing: Expressed Sequence Tags and Human Genome Project." *Science* 252 (5013): 1651–56. https://doi.org/10.1126/science.2047873.

Ali, Mohsin, Humna Hasan, Khola Rafique, Fakiha Afzal, Ghulam Kubra, Rabia Amir, Kandeel Shafique, et al. 2020. "Drought-Responsive ESTs in Wheat." In, 169–76. https://doi.org/10.1016/B978-0-12-819527-7.00010-8.

Alwine, J. C., D. J. Kemp, and G. R. Stark. 1977. "Method for Detection of Specific RNAs in Agarose Gels by Transfer to Diazobenzyloxymethyl-Paper and Hybridization with DNA Probes." *Proceedings of the National Academy of Sciences* 74 (12): 5350–54. https://doi.org/10.1073/pnas.74.12.5350.

Anders, Simon, Paul Theodor Pyl, and Wolfgang Huber. 2015. "HTSeq—a Python Framework to Work with High-Throughput Sequencing Data." *Bioinformatics* 31 (2): 166–69. https://doi.org/10.1093/bioinformatics/btu638.

Anders, Simon, Alejandro Reyes, and Wolfgang Huber. 2012. "Detecting Differential Usage of Exons from RNA-Seq Data." *Genome Research* 22 (10): 2008–17. https://doi.org/10.1101/gr.133744.111.

Bolger, Anthony M., Marc Lohse, and Bjoern Usadel. 2014. "Trimmomatic: A Flexible Trimmer for Illumina Sequence Data." *Bioinformatics*. https://doi.org/10.1093/bioinformatics/btu170.

Bolger, Anthony M., Marc Lohse, Bjoern Usadel, G K Sim, F C Kafatos, C W Jones, M D Koehler, et al. 2018. "De Novo Assembly and Analysis of RNA-Seq Data." *Nature Methods* 7 (1): 909–12. https://doi.org/10.1093/nar/gkm549.

Chatterjee, Aniruddha, Antonio Ahn, Euan J Rodger, Peter A Stockwell, and Michael R Eccles. 2018. "A Guide for Designing and Analyzing RNA-Seq Data." *Methods in Molecular Biology (Clifton, N.J.)* 1783: 35–80. https://doi.org/10.1007/978-1-4939-7834-2_3.

Chen, Shifu, Yanqing Zhou, Yaru Chen, and Jia Gu. 2018. "Fastp: An Ultra-Fast All-in-One FASTQ Preprocessor." *Bioinformatics (Oxford, England)* 34 (17): i884–90. https://doi.org/10.1093/bioinformatics/bty560.

Chen, Yuxin, Yongsheng Chen, Chunmei Shi, Zhibo Huang, Yong Zhang, Shengkang Li, Yan Li, et al. 2018. "SOAPnuke: A MapReduce Acceleration-Supported Software for Integrated Quality Control and Preprocessing of High-Throughput Sequencing Data." *GigaScience* 7 (1): 1–6. https://doi.org/10.1093/gigascience/gix120.

Cock, Peter J A, Christopher J Fields, Naohisa Goto, Michael L Heuer, and Peter M Rice. 2010. "The Sanger FASTQ File Format for Sequences with Quality Scores, and the Solexa/Illumina FASTQ Variants." *Nucleic Acids Research* 38 (6): 1767–71. https://doi.org/10.1093/nar/gkp1137.

Conesa, Ana, Pedro Madrigal, Sonia Tarazona, David Gomez-Cabrero, Alejandra Cervera, Andrew McPherson, Michał Wojciech Szcześniak, et al. 2016. "A Survey of Best Practices for RNA-Seq Data Analysis." *Genome Biology* 17 (January): 13. https://doi.org/10.1186/s13059-016-0881-8.

Dobin, Alexander, Carrie A. Davis, Felix Schlesinger, Jorg Drenkow, Chris Zaleski, Sonali Jha, Philippe Batut, Mark Chaisson, and Thomas R. Gingeras. 2013. "STAR: Ultrafast Universal RNA-Seq Aligner." *Bioinformatics* 29 (1): 15–21. https://doi.org/10.1093/bioinformatics/bts635.

Dündar, Friederike, Luce Skrabanek, and Paul Zumbo. 2018. "Introduction to Differential Gene Expression Analysis." *Applied Bioinformatics Core*, no. September 2015: 1–86. http://chagall.med.cornell.edu/RNASEQcourse/Intro2RNAseq.pdf.

Finkler, Marc, and Albrecht Ott. 2019. "Bead-Based Assay for Spatiotemporal Gene Expression Control in Cell-Free Transcription-Translation Systems." *BioTechniques* 66 (1): 29–33. https://doi.org/10.2144/btn-2018-0097.

Goodwin, Sara, James Gurtowski, Scott Ethe-Sayers, Panchajanya Deshpande, Michael Schatz, and W Richard McCombie. 2015. "Oxford Nanopore Sequencing and de Novo Assembly of a Eukaryotic Genome." *BioRxiv*, 13490. https://doi.org/10.1101/013490.

Grabherr, Manfred G, Brian J Haas, Moran Yassour, Joshua Z Levin, Dawn A Thompson, Ido Amit, Xian Adiconis, et al. 2011. "Full-Length Transcriptome Assembly from RNA-Seq Data without a Reference Genome." *Nature Biotechnology* 29 (7): 644–52. https://doi.org/10.1038/nbt.1883.

Grada, Ayman, and Kate Weinbrecht. 2013. "Next-Generation Sequencing: Methodology and Application." *The Journal of Investigative Dermatology* 133 (8): e11. https://doi.org/10.1038/jid.2013.248.

Heller, Michael J. 2002. "DNA Microarray Technology: Devices, Systems, and Applications." *Annual Review of Biomedical Engineering* 4 (1): 129–53. https://doi.org/10.1146/annurev.bioeng.4.020702.153438.

Hurgobin, Bhavna. 2016. "Short Read Alignment Using SOAP2." *Methods in Molecular Biology (Clifton, N.J.)* 1374: 241–52. https://doi.org/10.1007/978-1-4939-3167-5_13.

Kim, Daehwan, Ben Langmead, and Steven L Salzberg. 2015. "HISAT: A Fast Spliced Aligner with Low Memory Requirements." *Nature Methods* 12 (4): 357–60. https://doi.org/10.1038/nmeth.3317.

Kim, Daehwan, Geo Pertea, Cole Trapnell, Harold Pimentel, Ryan Kelley, and Steven L Salzberg. 2013. "TopHat2: Accurate Alignment of Transcriptomes in the Presence of Insertions, Deletions and Gene Fusions." *Genome Biology* 14 (4): R36. https://doi.org/10.1186/gb-2013-14-4-r36.

Kodzius, Rimantas, Miki Kojima, Hiromi Nishiyori, Mari Nakamura, Shiro Fukuda, Michihira Tagami, Daisuke Sasaki, et al. 2006. "CAGE: Cap Analysis of Gene Expression." *Nature Methods* 3 (3): 211–22. https://doi.org/10.1038/nmeth0306-211.

Kristensen, Lasse S, Maria S Andersen, Lotte V W Stagsted, Karoline K Ebbesen, Thomas B Hansen, and Jørgen Kjems. 2019. "The Biogenesis, Biology and Characterization of Circular RNAs." *Nature Reviews Genetics* 20 (11): 675–91. https://doi.org/10.1038/s41576-019-0158-7.

Kulkarni, Meghana M. 2011. "Digital Multiplexed Gene Expression Analysis Using the Nanostring Ncounter System." *Current Protocols in Molecular Biology*, no. SUPPL.94: 1–17. https://doi.org/10.1002/0471142727.mb25b10s94.

Langmead, Ben, Cole Trapnell, Mihai Pop, and Steven L Salzberg. 2009. "Ultrafast and Memory-Efficient Alignment of Short DNA Sequences to the Human Genome." *Genome Biology* 10 (3): R25. https://doi.org/10.1186/gb-2009-10-3-r25.

Leng, Ning, John A Dawson, James A Thomson, Victor Ruotti, Anna I Rissman, Bart M G Smits, Jill D Haag, Michael N Gould, Ron M Stewart, and Christina Kendziorski. 2013. "EBSeq: An Empirical Bayes Hierarchical Model for Inference in RNA-Seq Experiments." *Bioinformatics (Oxford, England)* 29 (8): 1035–43. https://doi.org/10.1093/bioinformatics/btt087.

Li, Bo, and Colin N Dewey. 2011. "RSEM: Accurate Transcript Quantification from RNA-Seq Data with or without a Reference Genome." *BMC Bioinformatics* 12 (August): 323. https://doi.org/10.1186/1471-2105-12-323.

Li, Chunbo, Luopei Guo, Shengli Li, and Keqin Hua. 2021. "Single-Cell Transcriptomics Reveals the Landscape of Intra-Tumoral Heterogeneity and Transcriptional Activities of ECs in CC." *Molecular Therapy - Nucleic Acids* 24 (June): 682–94. https://doi.org/10.1016/j.omtn.2021.03.017.

Li, Heng, and Richard Durbin. 2009. "Fast and Accurate Short Read Alignment with Burrows-Wheeler Transform." *Bioinformatics (Oxford, England)* 25 (14): 1754–60. https://doi.org/10.1093/bioinformatics/btp324.

Liao, Yang, Gordon K Smyth, and Wei Shi. 2014. "FeatureCounts: An Efficient General Purpose Program for Assigning Sequence Reads to Genomic Features." *Bioinformatics (Oxford, England)* 30 (7): 923–30. https://doi.org/10.1093/bioinformatics/btt656.

Love, Michael I, Wolfgang Huber, and Simon Anders. 2014. "Moderated Estimation of Fold Change and Dispersion for RNA-Seq Data with DESeq2." *Genome Biology* 15 (12): 550. https://doi.org/10.1186/s13059-014-0550-8.

Lowe, Rohan, Neil Shirley, Mark Bleackley, Stephen Dolan, and Thomas Shafee. 2017. "Transcriptomics Technologies." *PLoS Computational Biology* 13 (5): 1–23. https://doi.org/10.1371/journal.pcbi.1005457.

MA, Marra, Hillier L, and Waterston RH. 1998. "Expressed Sequence Tags - ESTablishing Bridges between Genomes." *Trends in Genetics* 14: 4–7. https://doi.org/10.1016/S0168-9525(97)01355-3.

Matsumura, Hideo, Khairun Hisam Bin Nasir, Kentaro Yoshida, Akiko Ito, Günter Kahl, Detlev H Krüger, and Ryohei Terauchi. 2006. "SuperSAGE Array: The Direct Use of 26-Base-Pair Transcript Tags in Oligonucleotide Arrays." *Nature Methods* 3 (6): 469–74. https://doi.org/10.1038/nmeth882.

Matsumura, Hideo, Stefanie Reich, Akiko Ito, Hiromasa Saitoh, Sophien Kamoun, Peter Winter, Gunter Kahl, Monika Reuter, Detlev H Kruger, and Ryohei Terauchi. 2003. "Gene Expression Analysis of Plant Host-Pathogen Interactions by SuperSAGE." *Proceedings of the National Academy of Sciences of the United States of America* 100 (26): 15718–23. https://doi.org/10.1073/pnas.2536670100.

Meyers, Blake C, Tam H Vu, Shivakundan Singh Tej, Hassan Ghazal, Marta Matvienko, Vikas Agrawal, Jianchang Ning, and Christian D Haudenschild. 2004. "Analysis of the Transcriptional Complexity of Arabidopsis Thaliana by Massively Parallel Signature Sequencing." *Nature Biotechnology* 22 (8): 1006–11. https://doi.org/10.1038/nbt992.

Naiyer, S., A. Bhattacharya, and S. Bhattacharya. 2019. "Advances in Entamoeba Histolytica Biology through Transcriptomic Analysis." *Frontiers in Microbiology* 10 (AUG). https://doi.org/10.3389/fmicb.2019.01921.

Naiyer, Sarah, Devinder Kaur, Jamaluddin Ahamad, Shashi Shekhar Singh, Yatendra Pratap Singh, Vivek Thakur, Alok Bhattacharya, and Sudha Bhattacharya. 2019. "Transcriptomic Analysis Reveals Novel Downstream Regulatory Motifs and Highly Transcribed Virulence Factor Genes of Entamoeba Histolytica." *BMC Genomics* 20 (1): 206. https://doi.org/10.1186/s12864-019-5570-z.

Naiyer, Sarah, Shashi Shekhar Singh, Devinder Kaur, Amartya Mukherjee, Yatendra Pratap Singh, Alok Bhattacharya, and Sudha Bhattacharya. 2022. "Transcriptomic Analysis of Ribosome Biogenesis and Pre-RRNA Processing during Growth Stress in Entamoeba Histolytica." *Experimental Parasitology*, 108308. https://doi.org/10.1016/j.exppara.2022.108308.

Nelson, N J. 2001. "Microarrays Have Arrived: Gene Expression Tool Matures." *Journal of the National Cancer Institute*. United States. https://doi.org/10.1093/jnci/93.7.492.

Pan, Qun, Ofer Shai, Leo J Lee, Brendan J Frey, and Benjamin J Blencowe. 2008. "Deep Surveying of Alternative Splicing Complexity in the Human Transcriptome by High-Throughput Sequencing." *Nature Genetics* 40 (12): 1413–15. https://doi.org/10.1038/ng.259.

Park, Munsu, Dongin Lee, Duhee Bang, and Ji Hyun Lee. 2020. "MAPS-Seq: Magnetic Bead-Assisted Parallel Single-Cell Gene Expression Profiling." *Experimental & Molecular Medicine* 52 (5): 804–14. https://doi.org/10.1038/s12276-020-0433-x.

Parkinson, John, and Mark Blaxter. 2009. "Expressed Sequence Tags: An Overview." *Methods in Molecular Biology (Clifton, N.J.)* 533: 1–12. https://doi.org/10.1007/978-1-60327-136-3_1.

Patel, Ravi K, and Mukesh Jain. 2012. "NGS QC Toolkit: A Toolkit for Quality Control of next Generation Sequencing Data." *PloS One* 7 (2): e30619. https://doi.org/10.1371/journal.pone.0030619.

Pertea, Mihaela, Geo M. Pertea, Corina M. Antonescu, Tsung Cheng Chang, Joshua T. Mendell, and Steven L. Salzberg. 2015. "StringTie Enables Improved Reconstruction of a Transcriptome from RNA-Seq Reads." *Nature Biotechnology* 33 (3): 290–95. https://doi.org/10.1038/nbt.3122.

Pozhitkov, Alex E., Diethard Tautz, and Peter A. Noble. 2007. "Oligonucleotide Microarrays: Widely Applied - Poorly Understood." *Briefings in Functional Genomics and Proteomics* 6 (2): 141–48. https://doi.org/10.1093/bfgp/elm014.

Roberts, Adam, and Lior Pachter. 2013. "Streaming Fragment Assignment for Real-Time Analysis of Sequencing Experiments." *Nature Methods* 10 (1): 71–73. https://doi.org/10.1038/nmeth.2251.

Robertson, Gordon, Jacqueline Schein, Readman Chiu, Richard Corbett, Matthew Field, Shaun D Jackman, Karen Mungall, et al. 2010. "De Novo Assembly and Analysis of RNA-Seq Data." *Nature Methods* 7 (11): 909–12. https://doi.org/10.1038/nmeth.1517.

Robinson, Mark D, Davis J McCarthy, and Gordon K Smyth. 2010. "EdgeR: A Bioconductor Package for Differential Expression Analysis of Digital Gene Expression Data." *Bioinformatics* 26 (1): 139–40. https://doi.org/10.1093/bioinformatics/btp616.

Schena, M, D Shalon, R W Davis, and P O Brown. 1995. "Quantitative Monitoring of Gene Expression Patterns with a Complementary DNA Microarray." *Science (New York, N.Y.)* 270 (5235): 467–70. https://doi.org/10.1126/science.270.5235.467.

Schmieder, Robert, and Robert Edwards. 2011. "Quality Control and Preprocessing of Metagenomic Datasets." *Bioinformatics (Oxford, England)* 27 (6): 863–64. https://doi.org/10.1093/bioinformatics/btr026.

Shiraki, T., S. Kondo, S. Katayama, K. Waki, T. Kasukawa, H. Kawaji, R. Kodzius, et al. 2003. "Cap Analysis Gene Expression for High-Throughput Analysis of Transcriptional Starting Point and Identification of Promoter Usage." *Proceedings of the National Academy of Sciences* 100 (26): 15776–81. https://doi.org/10.1073/pnas.2136655100.

Sim, G K, F C Kafatos, C W Jones, M D Koehler, A Efstratiadis, and T Maniatis. 1979. "Use of a CDNA Library for Studies on Evolution and Developmental Expression of the Chorion Multigene Families." *Cell* 18 (4): 1303–16. https://doi.org/10.1016/0092-8674(79)90241-1.

Soverini, Simona, Elisabetta Abruzzese, Monica Bocchia, Massimiliano Bonifacio, Sara Galimberti, Antonella Gozzini, Alessandra Iurlo, et al. 2019. "Next-Generation Sequencing for BCR-ABL1 Kinase Domain Mutation Testing in Patients with Chronic Myeloid Leukemia: A Position Paper." *Journal of Hematology & Oncology* 12 (1): 131. https://doi.org/10.1186/s13045-019-0815-5.

Sutcliffe, J G, R J Milner, F E Bloom, and R A Lerner. 1982. "Common 82-Nucleotide Sequence Unique to Brain RNA." *Proceedings of the National Academy of Sciences of the United States of America* 79 (16): 4942–46. https://doi.org/10.1073/pnas.79.16.4942.

Tang, Min, Ling Kui, Guanyi Lu, and Wenqiang Chen. 2020. "Disease-Associated Circular RNAs: From Biology to Computational Identification." *BioMed Research International* 2020. https://doi.org/10.1155/2020/6798590.

Trapnell, Cole, Brian A Williams, Geo Pertea, Ali Mortazavi, Gordon Kwan, Marijke J van Baren, Steven L Salzberg, Barbara J Wold, and Lior Pachter. 2010. "Transcript Assembly and Quantification by RNA-Seq Reveals Unannotated Transcripts and Isoform Switching during Cell Differentiation." *Nature Biotechnology* 28 (5): 511–15. https://doi.org/10.1038/nbt.1621.

Velculescu, V.E., L. Zhang, B. Vogelstein, and Kenneth W Kinzler. 2009. "Serial Analysis of Gene Expression." *Science* 270 (5235): 484–87. https://doi.org/10.1126/science.270.5235.484.

Wang, Zhong, Mark Gerstein, and Michael Snyder. 2009. "Nrg2484-1." *NATURE REVIEwS | Genetics* VOLUME 10 (jANUARy 2009): 57–63.

Wu, Jianzhong, Tomoko Maehara, Takanori Shimokawa, Shinichi Yamamoto, Chizuko Harada, Yuka Takazaki, Nozomi Ono, et al. 2002. "A Comprehensive Rice Transcript Map Containing 6591 Expressed Sequence Tag Sites." *The Plant Cell* 14 (3): 525–35. https://doi.org/10.1105/tpc.010274.

Yin, Huabing, and Damian Marshall. 2012. "Microfluidics for Single Cell Analysis." *Current Opinion in Biotechnology* 23 (1): 110–19. https://doi.org/10.1016/j.copbio.2011.11.002.

Chapter 2

Transcriptomics in Understanding the Basic Biology of the Cell

Mrigya Babuta[1,*] and Kunwar Somesh Vikramdeo[2,†]

[1]Beth Israel Deaconess Medical Center, Harvard Medical School, Boston, MA, USA

[2]Mitchell Cancer Institute, University of South Alabama, Mobile, AL, USA

Abstract

The main role played by DNA in the context of cells is the transferring of genetic information to daughter cells. While RNA can also act as an information carrier, its importance lies in the decoding of the information in the DNA and thus it is indispensable for normal function and growth of the cell. This property of RNA is fundamental to the process of the central dogma which states that information encoded in DNA is translated into proteins through mRNA. Although it seems straightforward, the variation in the stability and degradation rates of both RNA and proteins often result in differences in mRNA to protein expression in a time dependent and tissue specific manner. While the term transcriptome classically included mRNA, the template for protein synthesis, more than 95% of the cellular RNA is not translated into protein. However, these non-coding RNA transcripts have been well documented as key players in regulating the expression of many important signaling pathways and contribute towards maintenance of cellular homeostasis. Therefore, the term transcriptomics has since been amended to include all the RNA present in the cells including all forms of non-coding RNA.

*Corresponding Author's Email: mrigyababuta17@gmail.com.

†Corresponding Author's Email: somesh.jnu@gmail.com.

In: Transcriptomics and their Importance

Editors: Sarah Naiyer and. Khalid Hussain Bhat

ISBN: 979-8-88697-044-9

Increasing evidence highlights the role of transcriptomics in identifying the function of genes from non-model organisms or the organisms from which the reference genome is still not available. Further, its role is inevitable in identifying the genes and signaling networks which undergo changes upon countering different environmental stress conditions. Transcriptome profiling provides us with valuable information such as identifying biomarkers for a disease, distinction between different stages or subtypes of disease and helps to identify the correlation between genetic profile and expression pattern in healthy vs diseased individuals. Transcriptomic profiling provides a valuable tool to identify fusion transcripts as well as novel transcripts especially in the context of carcinogenesis and thus could be beneficial in devising a personalized cancer therapeutic approach. Furthermore, transcriptome analysis can also provide a useful perspective on the functional genomics of neurological disorders as their etiology is otherwise very complex on account of their propensity to be influenced by various genetic and environmental factors.

Transcriptomics has a vast application in different aspects of cellular function and a closer examination of the applications will amplify our knowledge about the finely tuned genetic functions and how they modulate the signaling pathways and have biological implications. Over the last few decades, the advent of newer technologies has redefined the scope of investigation and its concurrent use with other omics approaches could provide us with a bigger understanding of complex cellular processes. Current knowledge on the role of transcriptomics in elucidating the cellular and molecular functions is summarized in this chapter.

Keywords: transcriptomics, RNA-sequencing, gene annotation, host-pathogen interaction, neurooncology, biomarker

1. Introduction

The term "transcriptome" encompasses the total number and quantity of RNA transcripts in the cell. Transcriptome can vary depending upon different factors such as cell type and lineage, developmental stage and/or different physiological conditions. Thus, information about transcriptomes essentially provides us with key fundamental information about the functional and molecular elements of the genome and molecular constituents under different conditions.

Transcriptomics comprises the cataloguing of all the RNA transcripts present in the cell under a particular developmental stage or physiological stage (Lowe et al. 2017; Pan et al. 2008). It also includes elucidation of transcriptional structure of the genes Vis a Vis its transcription start and end sites, splicing patterns & post transcriptional modifications (phosphorylation, acetylation, etc.) (Pan et al. 2008).

Various techniques have been used over the years to decipher the transcriptome which can be broadly divided into two categories; first include hybridization-based strategies in which fluorescently labeled cDNA is incubated with customized microarrays (Figure1). Hybridization based methods have the advantage of being cheaper and can map the genome with high resolution. However, this method is limited by its dependency on existing information about the genome and scant range of detection. In stark contrast, sequence-based methods are much more precise owing to direct measurement of cDNA sequences.

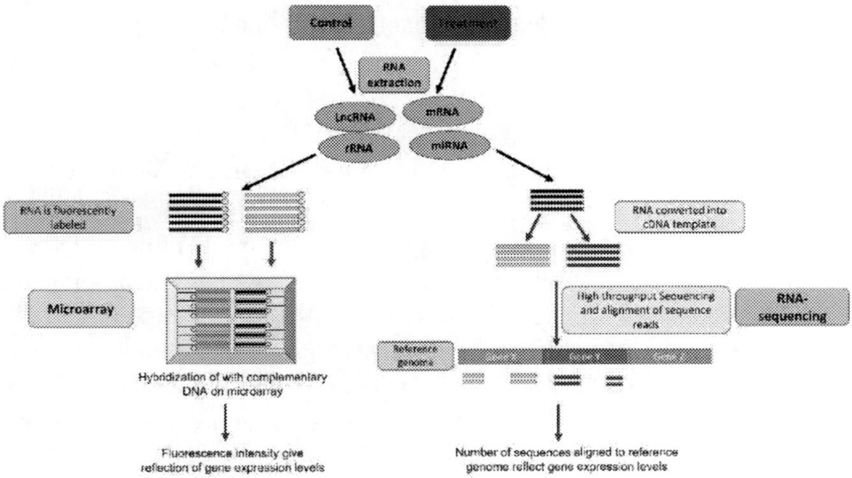

Figure 1. Schematic of transcriptomics techniques. Gene expression can be analysed either using Microarray based method in which RNA is fluorescently labelled or by using RNA sequencing which is high-throughput technique involving conversion of RNA into cDNA.

Traditionally, Sanger sequencing-based approaches were used to generate precise and accurate maps of gene expression levels (Gomes and Korf 2018). However, these methods often analyzed only a portion of the transcript and were unable to differentiate between isoforms of the gene.

Fortunately, the advent of high throughput sequencing methods has provided an answer to these glaring issues. One of the main challenges in understanding expression of gene regulation and molecular function of cells is the variable response of the same set of genes towards different tissue types, environment, and developmental stages. Transcriptomics, bolstered with extensive growth in sequencing techniques, has emerged as a boon in context of finding a solution for this challenge, giving us valuable insight on gene structure, function and regulation (Casamassimi et al. 2017; McGettigan 2013b) RNA seq is one of the recently developed transcriptomics technique which utilizes a "high throughput approach' such as NGS to identify each and every RNA transcript present inside a particular sample. (McGettigan 2013a; Z. Wang, Gerstein, and Snyder 2009). The most important aspect of this technology is the pace at which it has enhanced our knowledge about the variability of gene expression in different cell and tissue types and has elaborated the role of alternative splicing in this context. This has led to discovery of additional novel functional and regulatory elements of human genome such as single-nucleotide variants (SNPs) and novel gene fusion products in cancer.

2. Different RNA Species and Their Significance within Cell

The central dogma has underlined the significance of RNA in protein synthesis through transcription into mRNA and the further translation into protein with the help of ribosomes which are made up of rRNA. It is well established fact that in higher eukaryotes, a very little percentage of genome codes for proteins. The majority of non-coding portion of genome was historically classified as junk DNA (Palazzo and Gregory 2014). However, after decades of research, this perception about the non-coding part of genome has now found to be untrue.Non-coding RNAs are now, irrefutably considered as key gene regulatory molecules. The various RNA species in the cells can be broadly divided into two categories:

2.1. Coding Transcripts

The coding transcript in the cell is referred to as messenger RNA (mRNA) which is transcribed from the DNA via the action RNA polymerase. They include the information for the synthesis of proteins which is translated with the help of two other RNAs. First is the tRNA, which is the important molecules in context of reading the codon on the mRNA and possesses the corresponding amino acid. The second is rRNA, which makes up the ribosome, which moves along the mRNA and perform the synthesis of polypeptide chain. While tRNA and rRNA do not code for proteins per se, their role in the cell is exclusively reserved for the synthesis of proteins. mRNA, owing to its indispensability in the protein synthesis, can provide a pretty accurate picture of metabolic and molecular status of the cell under a particular set of conditions. The analysis of mRNAs, especially when coupled with whole genome sequencing is a valuable tool to assess the gene regulation network with robust precision. The various important bits of information gleaned from mRNA analysis include identification of different splicing forms of transcripts correlating with a variety of protein isoforms. It can also help us to ascertain the differential role played by the genotype of an organism on its gene expression profile (Huang et al. 2020).

2.2. Non-Coding Transcripts

A major fraction of genetic material materials in higher animals does not code for proteins. Although this part of the genome was earlier considered to be non functional, numerous studies in the recent past have shown that major part of this non coding genome is actively transcribed, their expression under tight regulation, and they play an incredibly significant role in the cells (ENCODE Project Consortium 2012; Hon et al. 2017).

Non-coding transcripts included RNA species other than mRNA such as tRNAs, rRNAs, miRNAs, lncRNA and piwiRNAs. Except for tRNA and rRNAs, which are crucial for protein synthesis, all other non-coding RNAs primarily play a regulatory role in the cells.

2.2.1. Long Non-Coding RNAs

Long non-coding RNAs (lncRNAs) are long RNA nucleotides, more than 200, not transcribed into functional proteins. lncRNAs vary greatly according to their biogenesis and genomics. It has been estimated that human genome codes for more than 100,000 lncRNAs, transcribed by variety of RNA polymerases. LncRNAs are known to share the methylation and polyadenylation pattern with mRNAs. LncRNAs play an important role in regulating gene expression. LncRNAs can initiate chromatin remodeling by recruiting a variety of factors.

In addition, they also known to be involved in initiation of transcription, play an important role in formations of organelles (Naganuma and Hirose 2013; Bonasio and Shiekhattar 2014). They also act as sponge for various miRNAs along with moderating mRNA expression and thus modulate gene expression at different levels, (Gong and Maquat 2011; Tay, Rinn, and Pandolfi 2014; Yoon, Abdelmohsen, and Gorospe 2014).

2.2.2. MicroRNAs

microRNAs are small RNAs about 22-25 nucleotides in length. Their significance in gene regulation was unraveled in *C. elegans.* Since then, their crucial and indispensable role in regulation of gene expression has been extensively elucidated in several other species including humans (He and Hannon 2004). miRNAs regulate gene expression by complementarily binding to 3'UTR of their target gene's mRNA and causing their degradation. Often the binding of miRNAs to the target mRNA is not perfect, in which case the inhibition of translation occurs (Humphreys et al. 2005; Valencia-Sanchez et al. 2006). They are, therefore, a major effector of post-transcriptional gene silencing.

2.2.3. Small Nuclear and Nucleolar RNAs

These are the small regulatory RNAs that are known to regulate gene expression through RNA splicing and processing the pre-mRNA. They display nuclear localization and are usually present in tightly bound complexes with other proteins, referred to as small nuclear ribonucleoproteins (snRNAs) (Matera, Terns, and Terns 2007). Small nucleolar RNAs (snoRNAs) are abundant in nucleolar extracts where they play a part in processing rRNA molecules. They can be categorized in two categories, the C/D box or H/ACA box families| (Matera, Terns, and Terns 2007).

2.2.4. Piwi-interacting RNAs (piRNA)

These are recently discovered RNA species with 21-35 nucleotide length. They have been found to act as silencers of transposons along with help fight viral infections and regulation of gene expression (Madison-Villar et al. 2016; Siomi et al. 2011) They have been also reported to maintain genome stability in germline cells and have a role in gametogenesis. piRNAs act by guiding PIWI proteins to degrade target mRNA, enhance heterochromatin formation by promoting methylation of DNA (Ozata et al. 2019). Their detection has been made possible by the rapid advances in RNA sequencing and their expression has been reported to be divergent across species.

There have been continued discoveries of newer forms of noncoding RNAs with novel role in the cells. The abundance of RNA species playing such a diverse and integral part in the maintenance of cells has lent credence to the idea of "RNA world," which states that the evolution of RNA preceded both DNA and proteins and may have compensated for the role of both the molecules. Discovery of ribozymes, RNA molecules possessing the capacity to catalysis, and the fact that RNA can code for proteins presents a viable proof for this hypothesis. The contribution of RNA to the cells is still being unraveled via the discovery of newer regulatory RNA molecules. Further studies are needed to confirm the verity of this theory.

3. Transcriptomics in Understanding the Molecular Function

One of the first conquests of the transcriptomics was the elucidation of transcriptomic signature of human skin. Various studies over the years have performed detailed transcriptional signatures associated with skin in response to various factors such as ultraviolet light, cytokines, toxins, vitamins and allergens (Blumenberg 2013; Santoro et al. 2018). This has led to development of SkinSig, a comprehensive resource detailing gene expression signatures of various types of cells present in the skin under different physiological conditions (Shih et al. 2017).

3.1. Annotation of Gene Function

Transcriptome analysis has played a pivotal role in discovering the gene functions and further identifying the specific phenotype related to genes. Analysis of RNA sequencing data over the years has helped in improving gene annotations by discovering novel gene transcripts, new alternative splicing isoforms, and re-annotations of genes from reference genome. RNA seq has helped in gene annotation not only for the organisms having well annotated genome/model organisms but also from the non-model organism from which information about reference genome sequencing is missing.

To identify the differentially expressed transcripts of the model organism from the transcriptomics data two strategies could be deployed. The analysis could be performed either genome-based i.e using genome and gene annotation file or transcriptome-based by using transcriptome sequences for example, MMSEQ and rSeq (Kalari et al. 2014; Turro et al. 2011). The commonality of each analysis includes primarily deciding the gene set from the available public database such as RefSeq, ENSBLE, etc which will decide the profiling capacity of the genes (O'Leary et al. 2016). In genome-based analysis RNA seq data is mapped to reference genome and further exon-exon splice junctions is mapped using spliced aligner tool. However, in a transcriptome-based approach RNA seq data is mapped to transcriptome sequences and does not require a spliced aligner tool rather it requires an unspliced aligner tool such as Bowtie or TOPHAT (Ghosh and Chan 2016).

To identify the unannotated transcripts using the reference genome, two different modes of transcriptome reconstruction are employed; genome-guided or *de novo* construction is performed. The initial steps of genome-guided transcriptome reconstructions are as mentioned above. After mapping the exon-exon splice junctions, the novel transcripts are delineated via omission of the previously known annotated transcripts. In contrast, *De novo* transcriptome reconstruction utilizes an assembler software tool such as Bridger, Trinity etc, to assemble RNA-seq reads. These reads are then assembled to the reference genome. Furthermore, the genes which are not annotated are considered as novel annotations.

Cucumber genome/transcriptome analysis is one of the examples of improved annotation of the coding transcripts using the *de novo* transcriptome reconstruction. Li et al. used RNA-Seq reads from different tissues and identified twenty-thousand protein-coding genes. Furthermore, the reassembled cucumber genome identified approximately 8,700 genes with various structural modifications, 5,285 genes of which were novel and were not previously annotated (Li et al. 2011). In another study, the *de novo* transcriptome analysis from two koalas from different geographical areas led to the identification of Koala retrovirus (KoRV) transcripts. This study is important to understand the molecular genetics of the koala and to further have greater insight into the evolution of marsupials (Hobbs et al. 2014).

De novo transcriptome reconstruction is also an important tool to help in assembling RNA-Seq reads for which there is no reference genome present (Grabherr et al. 2011). Howe et al. identified approximately 200,000 SNPs, in Douglas-fir using de novo transcriptome analysis for which no reference genome was available. This SNP database thus generated was beneficial to Douglas-fir breeding programs and enhanced our understanding of genetic variations and their correlation with climate change response (Howe et al. 2013). Furthermore, *de novo* transcriptome assembly from the various tissue of lobster which does not have reference genome, helped to identify genes involved neuronal behavior as neurotransmitters and modulators (McGrath et al. 2016).

One of the drawbacks of the de novo transcriptome reconstruction is that it cannot be applied to genes which have highly similar sequences and/or are repetitive sequences.

Identifying novel gene transcripts using RNA seq from both model and non-model organisms has acted as a bridge in understanding the biology of organisms. However, it is imperative to mind the accuracy and quality of novel transcripts identified from newer transcriptomics methods. The reason being that those transcripts may not be full length but rather gene fragments necessitate the need for stringent quality control methods for assessing and quality of novel genes/transcripts and for annotation of gene function.

3.2. Exploring Signaling Networks Using Transcriptomics

Signaling cascades facilitates cell-cell communication which in turn helps in modulating cellular function such as cell growth, metabolism, and division. These pathways are mainly regulated by different post-translational modifications of signaling moieties such as phosphorylation, ubiquitylation, isgylation and many more which in turn regulates the downstream effector molecules. In the early years of discovering the role of signaling cascade, the focus remained on identifying the importance of individual signaling pathways and to understand the dynamics of each signaling moiety in the respective pathway. However, little research focused on the global picture of how different signaling pathways are interconnected and how the crosstalk between different signaling pathways can modulate the output response pertaining to different stimuli. For example, NF-KB is a pleiotropic signaling moiety which can regulate multiple signaling pathways and can modulate the innate immune response in various diseases (Lane et al. 2017).

Till date, different signaling pathways have been identified and novel interconnections between signaling pathways have been deciphered. For example, ZBP1 mediated necroptosis which regulates both apoptosis and necrosis in response to viral infection was recently discovered (Kuriakose and Kanneganti 2018). However, one of the main challenges in understanding the complex network of signaling pathways is the inter- and intra-cellular heterogeneity in response to the stimulus. At the single-cell level, there is a distinct variability in signaling response depending upon an array of different factors such as type of the cell, the nature of stimuli, and other environmental conditions. The heterogeneity at cellular level in response to stimulus can primarily be classified as either genetic or non-genetic heterogeneity (Lun and Bodenmiller 2020).

Genetic heterogeneity in the signaling network is implied when there is accumulation of mutations which perturb the function of proteins and further inhibit the functionality of the signaling network. This phenomenon of accumulation of mutations not only brings in the inherited variability in a set of cells but also re-wires the dynamics of the signaling network. Genomic instability in the cancer cells leads to the accumulation of mutations. In a study by Creixell et al. 6 types of network-attacking mutations (NAMs), which were changes in the kinases and scaffolding proteins which led to network rewiring, and the genesis and extinction of phosphorylation sites were identified (Creixell et al. 2015). Furthermore, authors developed a computational platform (ReKINect) to identify NAMs and discovered accumulation of mutation in five ovarian cancer cell lines which causes a bias towards phospho-threonine signaling, weakening of phosphorylation motifs, and kinase-inactivating hotspots in cancer (Creixell et al. 2015; Abbas, Keaton, and Dutta 2013).

Non-genetic heterogeneity at the cellular level is not because of the changes at the genomic level but rather is contributed by both intrinsic and extrinsic factors; for example, epigenetic modifications in response to stimuli lead to different responses by different cell types. Intrinsic heterogeneity is an attribute of molecular machinery and stoichiometry of molecules present in different cell types. Extrinsic heterogeneity is contributed by factors that leads to ununiformed transcriptome and the proteome of cells. Variables such as concentration of a stimulus (identities of neighboring cells), extracellular matrix stiffness (Chaudhuri et al. 2014), local crowdedness and spatial constraints (Frechin et al. 2015; Stallaert et al. 2018), nutrient and oxygen gradients (Wellen and Thompson 2010), cell cycle and cell volume (Rapsomaniki et al. 2018) interactively provide heterogeneous signaling network responses in isogenic cells.

In normal state this heterogeneity plays a quintessential role in regulating the biological processes such as tissue development and cell proliferation, however during diseased conditions this intercellular heterogeneity adds a layer of complexity and determines the efficacy of different therapeutic interventions. Due to recent advances in the technology, now we could systematically dissect the heterogeneity in signaling networks and can further understand the role of signaling pathways. For instance, NF-kB dynamics in response to lipopolysaccharide and how it modulates the global transcriptional response was established using live-cell imaging approaches alongside with single-cell sequencing technologies in the same individual cell (Lane et al. 2017). Different techniques could be implied to understand the signaling network such as single-cell transcriptomics, epigenomics, and spatial transcriptomics.

Single-cell RNA sequencing (scRNA-seq) measures the transcriptome of individual cells and helps to identify the different cell populations and novel subpopulations within a given organ. scRNA-seq identifies the global transcriptome profile in response to the stimuli and in turn helps to deduce the signaling dynamics and the regulation of signaling network in response to stimulus. Till date, numerous studies have used this technique to identify the inter cell communications and to understand the dynamics of signaling network, for instance Skelly et.al., performed scRNA-seq of the murine cardiac non-myocyte and identified understudied cardiac cell populations, such as mural cells and glia. Further their study highlighted the presence of extensive intercellular communication networks in heart and suggested the presence of sexual dimorphism in gene expression in the heart (Skelly et al. 2018). Multiple other studies have identified the intercellular communication in different cell types such as paracrine signaling regulates the inflammatory response to the stimulus in mouse bone marrow derived dendritic cells (DCs) (Shalek et al. 2014), ligand-receptor interactions across all cell types in a tumor microenvironment to understand the tumor ecosystem and the variability which occurs across tumors in different mouse tumor models (Kumar et al. 2018). Recently, scRNA-seq is applied in understanding the epithelial-to-mesenchymal transition (EMT) which drives multiple biological pathway such as embryonic development and plays a crucial role in cancer initiation, progression, and metastasis (Deshmukh et al. 2021). Sc-RNAseq is also applied to understand the biology of COVID-19 infection and the host immune response. A study by Speranza et al. determined that pneumocytes are the site of viral replication, whereas macrophages were the drivers of inflammatory response (Speranza et al. 2021). scRNA-seq can be combined with other techniques such as single-cell epigenomics to understand the non-genetic heterogeneity in signaling network such as transcriptional regulation during stem cell differentiation (Guo et al. 2017).

Spatial transcriptomics helps to visualize and quantitatively analyze the transcriptome in histological tissue sections. Spatial transcriptomics identifies the transcripts using different methods such as fluorescent *in situ* sequencing (FISSEQ) (Lee et al. 2014), multiplexed MERFISH (K. H. Chen et al. 2015), and spatial barcoding (Ståhl et al. 2016). In FISSEQ, cDNA amplicons are stably crosslinked in the biological sample (Lee et al. 2014) whereas in spatial barcoding unique positional barcodes are used (Ståhl et al. 2016). Spatial transcriptomics helps in physically localizing novel gene sets which are uniquely expressed in specific cells. However, one of the major limitations of spatial transcriptomics is the in-depth transcriptomics information of cells in context of their localization within a tissue. Therefore, in recent time, multiple studies have integrated the spatial transcriptomics with scRNA-seq to enhance resolution in tissue (Longo et al. 2021). In cutaneous squamous cell carcinoma (cSCC) combination of spatial barcoding and scRNA-seq identified four tumor subpopulations, out of which three were similar to normal epidermal states, and fourth one was tumor-specific keratinocyte (TSK) population which was unique to cancer. TSK localized to a fibrovascular niche and act as a hub for intercellular communication (Ji et al. 2020). Similarly, in pancreatic ductal adenocarcinoma, combination of both the techniques identified the role of inflammatory fibroblasts in cancer stress responses (Moncada et al. 2020). A distinct subset of microglia has been implicated in the development of Alzheimer disease in mouse brains, and localized to amyloid plaques (Keren-Shaul et al. 2017). scRNA-seq when combined with epigenomic transcriptomics and spatial barcoding mapped the gene networks which is involved in differentiation of fibroblast to myofibroblast and further contributes towards the pathogenesis of myocardial infarction (Kuppe et al. 2020). A combination of different transcriptomics methods has opened great avenues in understanding the signaling network and dynamics.

4. Transcriptomics and Its Role in Understanding Cellular Function

Recent transcriptomics studies have uncovered a previously unheard phenomenon of transcription interference (TI) (Boldogkői, Tombácz, and Balázs 2019). TI is a mode of gene regulation that is characterized by the competition between transcriptional machinery of nearby and distant genes. TI is thought to facilitate the gene expression in a spatiotemporal manner via direct physical interaction between transcriptional machinery and may have evolved into another mode of regulation of gene expression along with conventional mode of transcription factor mediated regulation. In fact, by affecting transcription of genes even at distal positions within the genome, it may be a part a larger gene regulation network termed as transcription interference network and can work to suppress transcriptional noise generated by expression of irrelevant genes in context of a particular developmental stage of the cell and may also help in fine-tuning gene expression which otherwise would require use of cis- and trans-acting elements.

4.1. Understanding Host Pathogen Interactions

Host−pathogen interactions are dynamic information exchanges between both the partners which leads to activation of different molecular pathways. Predominantly, studies related to host-pathogen interaction is one-sided and lacks the global picture from both the organisms involved. RNA-seq is a powerful tool in understanding both the interaction between host-pathogen and their response towards each other. For instance, RNA-Seq along with other omics approaches, identified genes transcribing virulence factors of pathogenic bacteria such as *Neisseria meningitidis*, *Yersinia pestis*, *Mycobacterium tuberculosis*, and *Staphylococcus aureus* which could provide indispensable knowledge about etiology and pathogenicity of diseases caused by these organisms and could be instrumental in identification of new vaccine targets and drug designing (Wu, Wang, and Jennings 2008).

scRNA-seq is applied to both host and pathogen, a study by Avraham et al. used fluorescently labeled Salmonella and host sc-RNAseq, to characterize how the changes in gene expression modulates infection and infection phenotypes (Avraham et al. 2015). Host macrophage response after Salmonella infection identified a set of genes encoding for the type I IFN response, induced in one third of the infected cells. Furthermore, it was also observed that bacteria that express high levels of PhoPQ activity leads to higher expression of type I interferon when taken up host cells (Avraham et al. 2015). In another study, using scRNA-seq macrophage polarization during Salmonella infection was deciphered (Saliba et al. 2016). Macrophages which had nongrowing Salmonella showed phenotype of pro-inflammatory M1 macrophages whereas macrophages with growing Salmonella display polarized into anti-inflammatory M2 macrophages suggesting fast-growing Salmonella overcomes host defence by reprogramming macrophage polarization (Saliba et al. 2016). To overcome the limitations of scRNA-seq in understanding host response and to increase the depth of information obtained, scRNA-seq was coupled with photoactivatable fluorescent reporters known as NICHE-seq which helps to understand spatially the cellular and molecular composition of the infectious niches (Medaglia et al. 2017). Holosequencing is another approach which helps to identify the small-RNA and mRNA transcriptome at the same time from the single cell and to helps to understand the transcriptome landscape (Xiao et al. 2018).

scRNA-seq in bacteria has overcome multiple hurdles before being developed into a reality. Multiple technical hurdles such as isolation of bacterial RNA by mechanical disruption interfered with RNA stability and quality, presence of bacterial RNA in femtogram amount, lack of poly A tail on mRNA and highly labile bacterial mRNAs prevented from developing an efficient scRNA-seq technique. Recently, Multiple Annealing and Tailing-based Quantitative scRNA-seq (MATQ-seq) was used to profile, *S. Typhimurium* and *P. aeruginosa* cells (Imdahl et al. 2020). Prokaryotic Expression profiling by Tagging RNA In situ and sequencing (PETRI-seq), a novel and efficient approach was used by Blattman et al. to profile both Gram-negative and Gram-positive bacteria with excellent purity and minimal bias (Blattman et al. 2020). A recently developed high throughput technique, called microSPLiT (microbial split-pool ligation transcriptomics), has been also been useful in identifying heterogeneity in transcriptome of Gram-negative and Gram-positive bacteria and revealing gene expression related to prophage induction and metabolic pathways in *B. subtilis* cells (Kuchina et al. 2021).

RNA sequencing analysis in protists parasite has proven to be beneficial in understanding the gene expression pattern during different infectious stages. *Plasmodium falciparum* perpetuates its infection through non-replicating sexual stage called gametocytes. However, *P. falciparum* could also asexually replicate in red blood cells of host. A study by Poran et al., performed scRNA-seq of 18,000 single malaria cultures undergoing sexual commitment and identified various transcriptional factors and chromatin modifying enzymes facilitating the expression of genes involved in gametocyte development (Poran et al. 2017). Another study, using low throughput but higher coverage approach examined transcriptional dynamics of the *P. falciparum* during different blood stages and identified multigene families which are activated during different stages (Reid et al. 2018). RNA-seq analysis of *Leishmania amazonensis* wild type and arginase knock out promastigotes and axenic amastigotes identified transcripts which contributed approximately 60% of encoded hypothetical proteins and 443 un-annotated transcripts. Furthermore, this study highlighted the differential regulation of gene expression between different stages of *Leishmania* life cycle (Aoki et al. 2017). Spliced Leader RNA-seq (SL-seq) has also been applied to understand the differential gene expression pattern in promastigotes obtained either from the midgut of the carrier of Leishmania i.e., sand fly or for cultured promastigotes. This study helped to identify several genes that were upregulated in promastigotes isolated from sandfly and are important during metacyclogenesis, a process by which *Leishmania* is converted into the highly infective metacyclic promastigotes (Alcolea et al. 2018). High-throughput RNA sequencing is also applied to ancient protist parasite *Entamoeba histolytica* trophozoites. This study performed RNA-seq in trophozites which were in contact with different environment such as virulent trophozoites forming liver abscess, or virulent parasites invading the human colon explants and identified the factors which contributes towards aggravation of the disease (Weber et al. 2016). Another study exploited RNA-seq to identify the differentially expressed genes and their association with *E. histolytica* virulence potential (Naiyer et al. 2019; 2021) Further, authors identified high expression of virulence-related genes and low expression of some kinases and BspA family genes (Naiyer et al. 2019).

Dual RNA-seq, a recently developed RNA-sequencing technique, has been employed to assess the differential expression of a variety of genes during host-pathogen interaction and has enabled identification of novel signaling changes in context of host–pathogen interactions (Westermann et al. 2016; Westermann, Barquist, and Vogel 2017). Dual RNA-seq has been applied to multiple pathogens in cell culture infection model such *Salmonella* (Stapels et al. 2018), *E. coli* (Mavromatis et al. 2015), *M. tuberculosis* (Zimmermann et al.,). It been also applied to understand the real-time interaction between host-parasite such as *L. donovani* (Shadab et al. 2019) and *P. falciparum* (Bradwell et al. 2020).

scRNA-seq and Dual RNA-seq has paved a way to broaden our understanding on host-pathogen interaction. It will be intriguing to apply these approaches to understand the novel host pathways which protect them from pathogen invasion and would help in identifying the cellular pathways worth targeting for interventions and for drug development.

4.2. Disease Profiling and Pathogenesis

In recent times, transcriptomic profiling has provided a valuable tool to investigate and understand finer molecular details about various diseases especially ones where the limited information has remained a hurdle in devising an effective therapy. RNA sequencing has advanced our knowledge and revolutionized the era of transcriptome profiling by making it feasible to quantify specific expression of both gene and the alleles simultaneously in a single experiment. Comprehensive databases encompassing the information about novel genes and their isoforms, and fusion transcripts have been generated using RNA seq. Some of the examples of such databases are ENCODE, a database of regulatory elements in the genome (ENCODE Project Consortium 2012) and TCGA, which is a database containing information about variable expression of genes in different types of cancer (Tomczak, Czerwińska, and Wiznerowicz 2015). Such tools have been especially useful in understanding the pathology of different diseases.

Analyzing the transcriptomic profile of a patient's blood has been helpful in identification of novel biomarkers in a variety of disorders. Another important benefit of blood-based approaches is that they are non-invasive and can be quite sensitive. Transcriptomics will continue to provide new and better biomarkers for the various disorders. For example, immediate early response 3 gene (IER3) was identified as a potential biomarker for Huntington's disease (Runne et al. 2007). In another example, a variety of new and interesting biomarkers have been identified through RNA profiling of blood from patients suffering from melidiosis and could be helpful to devise a potential therapeutic approach for this lethal tropical disease (Yimthin et al. 2021).

Transcriptomics has clarified the finer workings of the brain activity at molecular levels which is otherwise tricky considering the different gene expression profiles of different regions of the brain. For example, in cases of disorders like HIV-associated neurocognitive disorder (HAND), whose therapy is plagued by lack of sufficient understanding, important information has been gleaned by RNA–seq pertaining to its pathogenesis. RNA seq analysis has pointed out the role of CCAAT enhancer binding protein (CEBPβ) in regulation of inflammation, neurogenesis, altered mitochondrial function and autophagy associated with HAND (Canchi et al. 2020). Continued investigation in this direction may provide information about the downstream pathways and help design more robust therapies in neurological abnormalities like HAND. Single cell transcriptomics has been used to generate enhanced transcriptomic profiles of human cortical neurons during different neurodevelopmental stages and was helpful in forming a comprehensive gene expression program of various developmental stages of cortical neurons (X. Wang and Cairns 2014). The information gleaned from such a profile could provide us with the knowledge about specific cell types and developmental stages leading to improper formation of specific cortical circuit's induction of disorders like autism and schizophrenia (Morabito et al. 2020).

Perhaps the most important contribution that transcriptomics has made is to enhance our knowledge in the field of cancer biology. RNA-seq especially is proving to be a powerful tool in characterizing the molecular landscape of malignant tumors of the central nervous system owing to its ability to quickly analyse the whole transcriptome (Cui et al. 2021; Krieger et al. 2020). In fact, its success has prompted its incorporation into conventional neuro-oncology treatment programs, where it can be used to gain crucial information about the tumor and thus can be effective in designing personalized targeted therapies.

Advent of microfluidics based sequencing platforms has been quite instrumental in enhancing the quality and accuracy of the data generated through RNA profiling of different type of tumors (Gierahn et al. 2017; Macosko et al. 2015; Klein et al. 2015). The only issue with droplet-based methods is the reduction in number and proportion of genes that can be analyzed at a given time. However, their cost effectiveness and ability to identify rare and distinct subpopulations within a tumor outweighs that shortcoming. It has uncovered some new features in cancer cells that were previously unseen. For example, single cell RNA seq discovered that the dominant class of cells in prostate cancer is basal cell type contrary to previous general notion that they are absent in prostate tumors (S. Chen et al. 2021). Single cell RNA seq combined with microfluidics platform has the potential to significantly advance the accuracy of scRNA-seq studies as it can help to reduce the limiting factors such as cell subtype-biasing, processing time, and/or stress response (Lombardo et al. 2021). Constant advancements in the field of transcriptomics and its integration with NGS and microfluidics platforms has provided us with considerable insight into the molecular profile of cancer cells, especially identification of several prognostic and predictive cancer biomarkers and has been instrumental in devising transcriptome-based precision medicine approach in clinics.

The above examples show how utilization of different approaches such as microarrays used in conjunction with the NGS platforms for RNA sequencing is advancing our knowledge and providing us better tools to combat various diseases. In fact, they have been instrumental in stratifying patients according to their molecular profile and devising a personalized treatment regimen for them. Over the course of last few decades, a variety of useful and innovative strategies have been employed in order to effectively identify and quantitate significant RNA species in a larger number of pathological conditions. Exciting new avenues are ripe for picking as transcriptomics integrated with other omics approaches, single cell and spatial omics, has immense potential for giving us a complete and comprehensive picture of alteration of cells and tissues in different pathological sates and helpful in precision medicine strategy.

Conclusion

Despite the valuable and humongous information provided by the transcriptomics approach, there are still caveats to using this technique. These include but are not limited to significant biological and technical difficulties in data integration which pose problems in merging datasets and limit our ability to observe potentially interesting results. The limitations include the following: a.) Discrepancy in nomenclatures of gene or proteins IDs across different databases or sources, b.) Variability in sample processing and use of different data source such as use of array or Next generation sequencing platforms; c.) dependence on computational power and capacity of the platforms used; and d.) Scarcity of robust and accurate prediction models with reliable ability to account for signaling changes in context of different cell types and their spatial organization in tissues and lastly, changes in developmental stages. (Guinney et al. 2015).

However, the most important limitation to using a transcriptomic approach is the fact that transcriptomic profile of the cells is often not identical to proteome of the cells which may plant a reasonable doubt in context of the exact molecular working of the cell and how additional factors may influence its phenotype (Gholami et al. 2013). Recent efforts have been made to address this issue by trying to visualize cells in a 3D model while incorporating temporospatial differences which may influence intracellular behavior.

Most glaring shortcomings of transcriptomics approaches can be appreciably counteracted by coupling their usage in conjunction with other types of omics approaches. RNA sequencing when integrated with multi-omics technology, further enhances our understanding on correlating single cell analysis with pathophysiological conditions. For instance, integrating transcriptome data with epigenomics data using single-cell methylome and transcriptome sequencing (scM&T-seq) helped to identified DNA methylome and transcriptome data. Assessment of RNA species along with simultaneous identification of open chromatin structure in a single cell has also recently made possible by a modification of single cell RNA seq technique, known as single-cell combinatorial indexing chromatin accessibility or sci-CAR. Furthermore, transcriptome data can also be integrated with proteome data using sequencing method such as cellular indexing of transcriptomes and epitopes by sequencing (CITE-seq), and RNA expression and protein sequencing assay (REAP-seq). These types of combinatorial approaches have become more meaningful in recent times and can usher in a new era of comprehensive and intuitive analysis of the transcriptome and how they regulate the various phenotypes associated with different developmental stages and disease.

References

Abbas, Tarek, Mignon A Keaton, and Anindya Dutta. 2013. "Genomic Instability in Cancer." *Cold Spring Harbor Perspectives in Biology* 5 (3): a012914. https://doi.org/10.1101/cshperspect.a012914.

Alcolea, Pedro J., Ana Alonso, Loren Baugh, Carolyn Paisie, Gowthaman Ramasamy, Aarthi Sekar, Aakash Sur, Maribel Jiménez, Ricardo Molina, Vicente Larraga, Peter J. Myler. 2018. "RNA-Seq Analysis Reveals Differences in Transcript Abundance between Cultured and Sand Fly-Derived Leishmania Infantum Promastigotes." *Parasitology International* 67 (4): 476–80. https://doi.org/10.1016/j.parint.2018.03.008.

Aoki, Juliana Ide, Sandra Marcia Muxel, Ricardo Andrade Zampieri, Maria Fernanda Laranjeira-Silva, Karl Erik Müller, Audun Helge Nerland, and Lucile Maria Floeter-Winter. 2017. "RNA-Seq Transcriptional Profiling of Leishmania Amazonensis Reveals an Arginase-Dependent Gene Expression Regulation." *PLoS Neglected Tropical Diseases* 11 (10): e0006026. https://doi.org/10.1371/journal.pntd.0006026.

Avraham, Roi, Nathan Haseley, Douglas Brown Cristina Penaranda, Humberto B Jijon John J Trombetta, Rahul Satija, Alex K Shalek, Ramnik J Xavier, Aviv Regev, Deborah T Hung. 2015. "Pathogen Cell-to-Cell Variability Drives Heterogeneity in Host Immune Responses." *Cell* 162 (6): 1309–21. https://doi.org/10.1016/j.cell.2015.08.027.

Blattman, Sydney B, Wenyan Jiang, Panos Oikonomou, and Saeed Tavazoie. 2020. "Prokaryotic Single-Cell RNA Sequencing by in Situ Combinatorial Indexing." *Nature Microbiology* 5 (10): 1192–1201. https://doi.org/10.1038/s41564-020-0729-6.

Blumenberg, Miroslav. 2013. "Skinomics: Past, Present and Future for Diagnostic Microarray Studies in Dermatology." *Expert Review of Molecular Diagnostics* 13 (8): 885–94. https://doi.org/10.1586/14737159.2013.846827.

Boldogkői, Zsolt, Dóra Tombácz, and Zsolt Balázs. 2019. "Interactions between the Transcription and Replication Machineries Regulate the RNA and DNA Synthesis in the Herpesviruses." *Virus Genes* 55 (3): 274–79. https://doi.org/10.1007/s11262-019-01643-5.

Bonasio, Roberto, and Ramin Shiekhattar. 2014. "Regulation of Transcription by Long Noncoding RNAs." *Annual Review of Genetics* 48: 433–55. https://doi.org/10.1146/annurev-genet-120213-092323.

Bradwell, Katie R, Drissa Coulibaly, Abdoulaye K Koné, Matthew B Laurens, Ahmadou Dembélé, Youssouf Tolo, Karim Traoré, Amadou Niangaly, Andrea A Berry, Bourema Kouriba, Christopher V Plowe, Ogobara K Doumbo, Kirsten E Lyke, Shannon Takala-Harrison, Mahamadou A Thera, Mark A Travassos, David Serre. 2020. "Host and Parasite Transcriptomic Changes upon Successive Plasmodium Falciparum Infections in Early Childhood." Edited by Paola Flórez de Sessions. *MSystems* 5 (4). https://doi.org/10.1128/mSystems.00116-20.

Canchi, Saranya, Mary K Swinton, Robert A Rissman, and Jerel Adam Fields. 2020. "Transcriptomic Analysis of Brain Tissues Identifies a Role for CCAAT Enhancer Binding Protein β in HIV-Associated Neurocognitive Disorder." *Journal of Neuroinflammation* 17 (1): 112. https://doi.org/10.1186/s12974-020-01781-w.

Casamassimi, Amelia, Antonio Federico, Monica Rienzo, Sabrina Esposito, and Alfredo Ciccodicola. 2017. "Transcriptome Profiling in Human Diseases: New Advances and Perspectives." *International Journal of Molecular Sciences* 18 (8). https://doi.org/10.3390/ijms18081652.

Chaudhuri, Ovijit, Sandeep T Koshy, Cristiana Branco da Cunha, Jae-Won Shin, Catia S Verbeke, Kimberly H Allison, and David J Mooney. 2014. "Extracellular Matrix Stiffness and Composition Jointly Regulate the Induction of Malignant Phenotypes in Mammary Epithelium." *Nature Materials* 13 (10): 970–78. https://doi.org/10.1038/nmat4009.

Chen, Kok Hao, Alistair N Boettiger, Jeffrey R Moffitt, Siyuan Wang, and Xiaowei Zhuang. 2015. "RNA Imaging. Spatially Resolved, Highly Multiplexed RNA Profiling in Single Cells." *Science (New York, N.Y.)* 348 (6233): aaa6090. https://doi.org/10.1126/science.aaa6090.

Chen, Sujun, Guanghui Zhu, Yue Yang, Fubo Wang, Yu-Tian Xiao, Na Zhang, Xiaojie Bian, Yasheng Zhu, Yongwei Yu, Fei Liu, Keqin Dong, Javier Mariscal, Yin Liu, Fraser Soares, Helen Loo Yau, Bo Zhang, Weidong Chen, Chao Wang, Dai Chen, Qinghua Guo, Zhengfang Yi, Mingyao Liu, Michael Fraser, Daniel D. De Carvalho, Paul C. Boutros, Dolores Di Vizio, Zhou Jiang, Theodorus van der Kwast, Alejandro Berlin, Song Wu, Jianhua Wang, Housheng Hansen He and Shancheng Ren. 2021. "Single-Cell Analysis Reveals Transcriptomic Remodellings in Distinct Cell Types That Contribute to Human Prostate Cancer Progression." *Nature Cell Biology* 23 (1): 87–98. https://doi.org/10.1038/s41556-020-00613-6.

Creixell, Pau, Erwin M Schoof, Craig D Simpson, James Longden, Chad J Miller, Hua Jane Lou, Lara Perryman, Thomas R Cox, Nevena Zivanovic, Antonio Palmeri, Agata Wesolowska-Andersen, Manuela Helmer-Citterich, Jesper Ferkinghoff-Borg, Hiroaki Itamochi, Bernd Bodenmiller, Janine T Erler Benjamin E Turk, Rune Linding. 2015. "Kinome-Wide Decoding of Network-Attacking Mutations Rewiring Cancer Signaling." *Cell* 163 (1): 202–17. https://doi.org/10.1016/j.cell.2015.08.056.

Cui, Yueli, Chao Li, Zhenhuan Jiang, Shu Zhang, Qingqing Li, Xixi Liu, Yuan Zhou, Runting Li, Liudong Wei, Lianwang Li, Qi Zhang, Lu Wen, Fuchou Tang, Dabiao Zhou. 2021. "Single-Cell Transcriptome and Genome Analyses of Pituitary Neuroendocrine Tumors." *Neuro-Oncology*, April. https://doi.org/10.1093/neuonc/noab102.

Deshmukh, Abhijeet P, Suhas V Vasaikar, Katarzyna Tomczak, Shubham Tripathi, Petra den Hollander, Emre Arslan, Priyanka Chakraborty, Rama Soundararajan, Mohit Kumar Jolly, Kunal Rai, Herbert Levine, Sendurai A Mani. 2021. "Identification of EMT Signaling Cross-Talk and Gene Regulatory Networks by Single-Cell RNA Sequencing." *Proceedings of the National Academy of Sciences of the United States of America* 118 (19). https://doi.org/10.1073/pnas.2102050118.

ENCODE Project Consortium. 2012. "An Integrated Encyclopedia of DNA Elements in the Human Genome." *Nature* 489 (7414): 57–74. https://doi.org/10.1038/nature 11247.

Frechin, Mathieu, Thomas Stoeger, Stephan Daetwyler, Charlotte Gehin, Nico Battich, Eva-Maria Damm, Lilli Stergiou, Howard Riezman, and Lucas Pelkmans. 2015. "Cell-Intrinsic Adaptation of Lipid Composition to Local Crowding Drives Social Behaviour." *Nature* 523 (7558): 88–91. https://doi.org/10.1038/nature14429.

Gholami, Amin Moghaddas, Hannes Hahne, Zhixiang Wu, Florian Johann Auer, Chen Meng, Mathias Wilhelm, and Bernhard Kuster. 2013. "Global Proteome Analysis of the NCI-60 Cell Line Panel." *Cell Reports* 4 (3): 609–20. https://doi.org/10.1016/j. celrep.2013.07.018.

Ghosh, Sreya, and Chon-Kit Kenneth Chan. 2016. "Analysis of RNA-Seq Data Using TopHat and Cufflinks." *Methods in Molecular Biology (Clifton, N.J.)* 1374: 339–61. https://doi.org/10.1007/978-1-4939-3167-5_18.

Gierahn, Todd M, Marc H Wadsworth, Travis K Hughes, Bryan D Bryson, Andrew Butler, Rahul Satija, Sarah Fortune, J Christopher Love, and Alex K Shalek. 2017. "Seq-Well: Portable, Low-Cost RNA Sequencing of Single Cells at High Throughput." *Nature Methods* 14 (4): 395–98. https://doi.org/10.1038/nmeth.4179.

Gomes, Alicia, and Bruce Korf. 2018. "Genetic Testing Techniques." In *Pediatric Cancer Genetics*, 47–64. Elsevier. https://doi.org/10.1016/B978-0-323-48555-5.00005-3.

Gong, Chenguang, and Lynne E Maquat. 2011. "LncRNAs Transactivate STAU1-Mediated MRNA Decay by Duplexing with 3' UTRs via Alu Elements." *Nature* 470 (7333): 284–88. https://doi.org/10.1038/nature09701.

Grabherr, Manfred G, Brian J Haas, Moran Yassour, Joshua Z Levin, Dawn A Thompson, Ido Amit, Xian Adiconis, Lin Fan, Raktima Raychowdhury, Qiandong Zeng, Zehua Chen, Evan Mauceli, Nir Hacohen, Andreas Gnirke, Nicholas Rhind, Federica di Palma, Bruce W Birren, Chad Nusbaum, Kerstin Lindblad-Toh, Nir Friedman, Aviv Regev. 2011. "Full-Length Transcriptome Assembly from RNA-Seq Data without a Reference Genome." *Nature Biotechnology* 29 (7): 644–52. https://doi.org/10.1038/ nbt.1883.

Guinney, Justin, Rodrigo Dienstmann, Xin Wang, Aurélien de Reyniès, Andreas Schlicker, Charlotte Soneson, Laetitia Marisa, Paul Roepman, Gift Nyamundanda, Paolo Angelino, Brian M Bot, Jeffrey S Morris, Iris M Simon, Sarah Gerster, Evelyn Fessler, Felipe De Sousa E Melo, Edoardo Missiaglia, Hena Ramay, David Barras, Krisztian Homicsko, Dipen Maru, Ganiraju C Manyam, Bradley Broom, Valerie Boige, Beatriz Perez-Villamil, Ted Laderas, Ramon Salazar, Joe W Gray, Douglas Hanahan, Josep Tabernero, Rene Bernards, Stephen H Friend, Pierre Laurent-Puig, Jan Paul Medema, Anguraj Sadanandam, Lodewyk Wessels, Mauro Delorenzi, Scott Kopetz, Louis

Vermeulen, Sabine Tejpar. 2015. "The Consensus Molecular Subtypes of Colorectal Cancer." *Nature Medicine* 21 (11): 1350–56. https://doi.org/10.1038/nm.3967.

Guo, Jingtao, Edward J Grow, Chongil Yi, Hana Mlcochova, Geoffrey J Maher, Cecilia Lindskog, Patrick J Murphy, Candice L Wike, Douglas T Carrell, Anne Goriely, James M Hotaling, Bradley R Cairns. 2017. "Chromatin and Single-Cell RNA-Seq Profiling Reveal Dynamic Signaling and Metabolic Transitions during Human Spermatogonial Stem Cell Development." *Cell Stem Cell* 21 (4): 533-546.e6. https://doi.org/10.1016/j.stem.2017.09.003.

He, Lin, and Gregory J Hannon. 2004. "MicroRNAs: Small RNAs with a Big Role in Gene Regulation." *Nature Reviews. Genetics* 5 (7): 522–31. https://doi.org/10.1038/nrg 1379.

Hobbs, Matthew, Ana Pavasovic, Andrew G King, Peter J Prentis, Mark DB Eldridge, Zhiliang Chen, Donald J Colgan, Adam Polkinghorne, Marc R Wilkins, Cheyne Flanagan, Amber Gillett, Jon Hanger, Rebecca N Johnson and Peter Timms. 2014. "A Transcriptome Resource for the Koala (Phascolarctos Cinereus): Insights into Koala Retrovirus Transcription and Sequence Diversity." BMC Genomics 15 (September): 786. https://doi.org/10.1186/1471-2164-15-786.

Hon, Chung-Chau, Jordan A Ramilowski, Jayson Harshbarger, Nicolas Bertin, Owen JL Rackham, Julian Gough, Elena Denisenko, Sebastian Schmeier, Thomas M Poulsen, Jessica Severin, Marina Lizio, Hideya Kawaji, Takeya Kasukawa, Masayoshi Itoh, A Maxwell Burroughs, Shohei Noma, Sarah Djebali, Tanvir Alam, Yulia A Medvedeva, Alison C Testa, Leonard Lipovich, Chi-Wai Yip, Imad Abugessaisa, Mickaël Mendez, Akira Hasegawa, Dave Tang, Timo Lassmann, Peter Heutink, Magda Babina, Christine A Wells, Soichi Kojima, Yukio Nakamura, Harukazu Suzuki, Carsten O Daub, Michiel JL de Hoon, Erik Arner, Yoshihide Hayashizaki, Piero Carninci, Alistair RR Forrest. 2017. "An Atlas of Human Long Non-Coding RNAs with Accurate 5' Ends." *Nature* 543 (7644): 199–204. https://doi.org/10.1038/nature21374.

Howe, Glenn T, Jianbin Yu, Brian Knaus, Richard Cronn, Scott Kolpak, Peter Dolan, W Walter Lorenz, and Jeffrey FD Dean. 2013. "A SNP Resource for Douglas-Fir: De Novo Transcriptome Assembly and SNP Detection and Validation." *BMC Genomics* 14 (February): 137. https://doi.org/10.1186/1471-2164-14-137.

Huang, Li, Xia Liu, Manish K Pandey, Xiaoping Ren, Haiwen Chen, Xiaomeng Xue, Nian Liu, Dongxin Huai, Yuning Chen, Xiaojing Zhou, Huaiyong Luo, Weigang Chen, Yong Lei, Kede Liu, Yingjie Xiao, Rajeev K Varshney, Boshou Liao, Huifang Jiang. 2020. "Genome-Wide Expression Quantitative Trait Locus Analysis in a Recombinant Inbred Line Population for Trait Dissection in Peanut." *Plant Biotechnology Journal* 18 (3): 779–90. https://doi.org/10.1111/pbi.13246.

Humphreys, David T, Belinda J Westman, David I K Martin, and Thomas Preiss. 2005. "MicroRNAs Control Translation Initiation by Inhibiting Eukaryotic Initiation Factor 4E/Cap and Poly(A) Tail Function." *Proceedings of the National Academy of Sciences of the United States of America* 102 (47): 16961–66. https://doi.org/10.1073/pnas. 0506482102.

Imdahl, Fabian, Ehsan Vafadarnejad, Christina Homberger, Antoine-Emmanuel Saliba, and Jörg Vogel. 2020. "Single-Cell RNA-Sequencing Reports Growth-Condition-Specific Global Transcriptomes of Individual Bacteria." *Nature Microbiology* 5 (10): 1202–6. https://doi.org/10.1038/s41564-020-0774-1.

Ji, Andrew L, Adam J Rubin, Kim Thrane, Sizun Jiang, David L Reynolds, Robin M Meyers, Margaret G Guo, Benson M George, Annelie Mollbrink, Joseph Bergenstråhle, Ludvig Larsson, Yunhao Bai, Bokai Zhu, Aparna Bhaduri, Jordan M Meyers, Xavier Rovira-Clavé, S Tyler Hollmig, Sumaira Z Aasi, Garry P Nolan, Joakim Lundeberg, Paul A Khavari. 2020. "Multimodal Analysis of Composition and Spatial Architecture in Human Squamous Cell Carcinoma." *Cell* 182 (2): 497-514.e22. https://doi.org/10.1016/j.cell.2020.05.039.

Kalari, Krishna R, Asha A Nair, Jaysheel D Bhavsar, Daniel R O'Brien, Jaime I Davila, Matthew A Bockol, Jinfu Nie, Xiaojia Tang, Saurabh Baheti, Jay B Doughty, Sumit Middha, Hugues Sicotte, Aubrey E Thompson, Yan W Asmann, Jean-Pierre A Kocher. 2014. "MAP-RSeq: Mayo Analysis Pipeline for RNA Sequencing." *BMC Bioinformatics* 15 (June): 224. https://doi.org/10.1186/1471-2105-15-224.

Keren-Shaul, Hadas, Amit Spinrad, Assaf Weiner, Orit Matcovitch-Natan, Raz Dvir-Szternfeld, Tyler K Ulland, Eyal David, Kuti Baruch, David Lara-Astaiso, Beata Toth, Shalev Itzkovitz, Marco Colonna, Michal Schwartz, Ido Amit. 2017. "A Unique Microglia Type Associated with Restricting Development of Alzheimer's Disease." *Cell* 169 (7): 1276-1290.e17. https://doi.org/10.1016/j.cell.2017.05.018.

Klein, Allon M, Linas Mazutis, Ilke Akartuna, Naren Tallapragada, Adrian Veres, Victor Li, Leonid Peshkin, David A Weitz, and Marc W Kirschner. 2015. "Droplet Barcoding for Single-Cell Transcriptomics Applied to Embryonic Stem Cells." *Cell* 161 (5): 1187–1201. https://doi.org/10.1016/j.cell.2015.04.044.

Krieger, Teresa G, Stephan M Tirier, Jeongbin Park, Katharina Jechow, Tanja Eisemann, Heike Peterziel, Peter Angel, Roland Eils, and Christian Conrad. 2020. "Modeling Glioblastoma Invasion Using Human Brain Organoids and Single-Cell Transcriptomics." *Neuro-Oncology* 22 (8): 1138–49. https://doi.org/10.1093/neuonc/noaa091.

Kuchina, Anna, Leandra M Brettner, Luana Paleologu, Charles M Roco, Alexander B Rosenberg, Alberto Carignano, Ryan Kibler, Matthew Hirano, R William DePaolo, and Georg Seelig. 2021. "Microbial Single-Cell RNA Sequencing by Split-Pool Barcoding." *Science (New York, N.Y.)* 371 (6531). https://doi.org/10.1126/science.aba5257.

Kumar, Manu P, Jinyan Du, Georgia Lagoudas, Yang Jiao, Andrew Sawyer, Daryl C Drummond, Douglas A Lauffenburger, and Andreas Raue. 2018. "Analysis of Single-Cell RNA-Seq Identifies Cell-Cell Communication Associated with Tumor Characteristics." *Cell Reports* 25 (6): 1458-1468.e4. https://doi.org/10.1016/j.celrep.2018.10.047.

Kuppe, Christoph, Ricardo Omar Ramirez Flores, Li Zhijian, Monica Toft Hannani, Jovan Tanevski, Maurice Halder, Mingbo Cheng, Susanne Ziegler, Xiaoting Zhang, Fabian Priester, Nadine Kaesler, Yaoxian Xu, Remco Hoogenboezem, Eric MJ Bindels, Rebekka Schneider-Kramann, Hendrik Milting, Ivan Gesteira Costa Filho, Julio Saez-Rodriguez and Rafael Kramann. 2020. "Spatial Multi-Omic Map of Human

Myocardial Infarction." *BioRxiv*, January, 2020.12.08.411686. https://doi.org/10.1101/2020.12.08.411686.

Kuriakose, Teneema, and Thirumala-Devi Kanneganti. 2018. "ZBP1: Innate Sensor Regulating Cell Death and Inflammation." *Trends in Immunology* 39 (2): 123–34. https://doi.org/10.1016/j.it.2017.11.002.

Lane, Keara, David Van Valen, Mialy M DeFelice, Derek N Macklin, Takamasa Kudo, Ariel Jaimovich, Ambrose Carr, Tobias Meyer, Dana Pe'er, Stéphane C Boutet, Markus W Covert. 2017. "Measuring Signaling and RNA-Seq in the Same Cell Links Gene Expression to Dynamic Patterns of NF-KB Activation." *Cell Systems* 4 (4): 458-469.e5. https://doi.org/10.1016/j.cels.2017.03.010.

Lee, Je Hyuk, Evan R Daugharthy, Jonathan Scheiman, Reza Kalhor, Joyce L Yang, Thomas C Ferrante, Richard Terry, Sauveur SF Jeanty, Chao Li, Ryoji Amamoto, Derek T Peters, Brian M Turczyk, Adam H Marblestone, Samuel A Inverso, Amy Bernard, Prashant Mali, Xavier Rios, John Aach, George M Church. 2014. "Highly Multiplexed Subcellular RNA Sequencing in Situ." *Science (New York, N.Y.)* 343 (6177): 1360–63. https://doi.org/10.1126/science.1250212.

Li, Zhen, Zhonghua Zhang, Pengcheng Yan, Sanwen Huang, Zhangjun Fei, and Kui Lin. 2011. "RNA-Seq Improves Annotation of Protein-Coding Genes in the Cucumber Genome." *BMC Genomics* 12 (November): 540. https://doi.org/10.1186/1471-2164-12-540.

Lombardo, Jeremy A, Marzieh Aliaghaei, Quy H Nguyen, Kai Kessenbrock, and Jered B Haun. 2021. "Microfluidic Platform Accelerates Tissue Processing into Single Cells for Molecular Analysis and Primary Culture Models." *Nature Communications* 12 (1): 2858. https://doi.org/10.1038/s41467-021-23238-1.

Longo, Sophia K, Margaret G Guo, Andrew L Ji, and Paul A Khavari. 2021. "Integrating Single-Cell and Spatial Transcriptomics to Elucidate Intercellular Tissue Dynamics." *Nature Reviews. Genetics*, June. https://doi.org/10.1038/s41576-021-00370-8.

Lowe, Rohan, Neil Shirley, Mark Bleackley, Stephen Dolan, and Thomas Shafee. 2017. "Transcriptomics Technologies." *PLoS Computational Biology* 13 (5): e1005457. https://doi.org/10.1371/journal.pcbi.1005457.

Lun, Xiao-Kang, and Bernd Bodenmiller. 2020. "Profiling Cell Signaling Networks at Single-Cell Resolution." *Molecular & Cellular Proteomics: MCP* 19 (5): 744–56. https://doi.org/10.1074/mcp.R119.001790.

Macosko, Evan Z, Anindita Basu, Rahul Satija, James Nemesh, Karthik Shekhar, Melissa Goldman, Itay Tirosh, Allison R Bialas, Nolan Kamitaki, Emily M Martersteck, John J Trombetta, David A Weitz, Joshua R Sanes, Alex K Shalek, Aviv Regev and Steven A McCarroll. 2015. "Highly Parallel Genome-Wide Expression Profiling of Individual Cells Using Nanoliter Droplets." *Cell* 161 (5): 1202–14. https://doi.org/10.1016/j.cell.2015.05.002.

Madison-Villar, MJ, Cheng Sun, Nelson C Lau, Matthew L Settles, and Rachel Lockridge Mueller. 2016. "Small RNAs from a Big Genome: The PiRNA Pathway and Transposable Elements in the Salamander Species Desmognathus Fuscus." *Journal of Molecular Evolution* 83 (3–4): 126–36. https://doi.org/10.1007/s00239-016-9759-3.

Matera, A Gregory, Rebecca M Terns, and Michael P Terns. 2007. "Non-Coding RNAs: Lessons from the Small Nuclear and Small Nucleolar RNAs." *Nature Reviews. Molecular Cell Biology* 8 (3): 209–20. https://doi.org/10.1038/nrm2124.

Mavromatis, Charalampos Harris, Nilesh J Bokil, Makrina Totsika, Asha Kakkanat, Kolja Schaale, Carlo V Cannistraci, Taewoo Ryu, Scott A Beatson, Glen C Ulett, Mark A Schembri, Matthew J Sweet, Timothy Ravasi. 2015. "The Co-Transcriptome of Uropathogenic Escherichia Coli-Infected Mouse Macrophages Reveals New Insights into Host-Pathogen Interactions." *Cellular Microbiology* 17 (5): 730–46. https://doi.org/10.1111/cmi.12397.

McGettigan, Paul A. 2013a. "Transcriptomics in the RNA-Seq Era." *Current Opinion in Chemical Biology* 17 (1): 4–11. https://doi.org/10.1016/j.cbpa.2012.12.008.

———. 2013b. "Transcriptomics in the RNA-Seq Era." *Current Opinion in Chemical Biology* 17 (1): 4–11. https://doi.org/10.1016/j.cbpa.2012.12.008.

McGrath, Lara Lewis, Steven V Vollmer, Stefan T Kaluziak, and Joseph Ayers. 2016. "De Novo Transcriptome Assembly for the Lobster Homarus Americanus and Characterization of Differential Gene Expression across Nervous System Tissues." *BMC Genomics* 17 (January): 63. https://doi.org/10.1186/s12864-016-2373-3.

Medaglia, Chiara, Amir Giladi, Liat Stoler-Barak, Marco De Giovanni, Tomer Meir Salame, Adi Biram, Eyal David, Hanjie Li, Matteo Iannacone, Ziv Shulman, Ido Amit. 2017. "Spatial Reconstruction of Immune Niches by Combining Photoactivatable Reporters and ScRNA-Seq." *Science (New York, N.Y.)* 358 (6370): 1622–26. https://doi.org/10.1126/science.aao4277.

Moncada, Reuben, Dalia Barkley, Florian Wagner, Marta Chiodin, Joseph C Devlin, Maayan Baron, Cristina H Hajdu, Diane M Simeone, and Itai Yanai. 2020. "Integrating Microarray-Based Spatial Transcriptomics and Single-Cell RNA-Seq Reveals Tissue Architecture in Pancreatic Ductal Adenocarcinomas." *Nature Biotechnology* 38 (3): 333–42. https://doi.org/10.1038/s41587-019-0392-8.

Morabito, Samuel, Emily Miyoshi, Neethu Michael, and Vivek Swarup. 2020. "Integrative Genomics Approach Identifies Conserved Transcriptomic Networks in Alzheimer's Disease." *Human Molecular Genetics* 29 (17): 2899–2919. https://doi.org/10.1093/hmg/ddaa182.

Naganuma, Takao, and Tetsuro Hirose. 2013. "Paraspeckle Formation during the Biogenesis of Long Non-Coding RNAs." *RNA Biology* 10 (3): 456–61. https://doi.org/10.4161/rna.23547.

Naiyer, Sarah, Devinder Kaur, Jamaluddin Ahamad, Shashi Shekhar Singh, Yatendra Pratap Singh, Vivek Thakur, Alok Bhattacharya, and Sudha Bhattacharya. 2019. "Transcriptomic Analysis Reveals Novel Downstream Regulatory Motifs and Highly Transcribed Virulence Factor Genes of Entamoeba Histolytica." *BMC Genomics* 20 (1): 206. https://doi.org/10.1186/s12864-019-5570-z.

Naiyer, Sarah, Shashi Shekhar Singh, Devinder Kaur, Yatendra Pratap Singh, Amartya Mukherjee, Alok Bhattacharya, and Sudha Bhattacharya. 2021. "Transcriptomic Analysis of Ribosome Biogenesis and Pre-RRNA Processing during Growth Stress in *Entamoeba histolytica*" *BioRxiv*, January, 2021.08.01.454488. https://doi.org/10.1101/2021.08.01.454488.

O'Leary, Nuala A, Mathew W Wright, J Rodney Brister, Stacy Ciufo, Diana Haddad, Rich McVeigh, Bhanu Rajput, Barbara Robbertse, Brian Smith-White, Danso Ako-Adjei, Alexander Astashyn, Azat Badretdin, Yiming Bao, Olga Blinkova, Vyacheslav Brover, Vyacheslav Chetvernin, Jinna Choi, Eric Cox, Olga Ermolaeva, Catherine M Farrell, Tamara Goldfarb, Tripti Gupta, Daniel Haft, Eneida Hatcher, Wratko Hlavina, Vinita S Joardar, Vamsi K Kodali, Wenjun Li, Donna Maglott, Patrick Masterson, Kelly M McGarvey, Michael R Murphy, Kathleen O'Neill, Shashikant Pujar, Sanjida H Rangwala, Daniel Rausch, Lillian D Riddick, Conrad Schoch, Andrei Shkeda, Susan S Storz, Hanzhen Sun, Francoise Thibaud-Nissen, Igor Tolstoy, Raymond E Tully, Anjana R Vatsan, Craig Wallin, David Webb, Wendy Wu, Melissa J Landrum, Avi Kimchi, Tatiana Tatusova, Michael DiCuccio, Paul Kitts, Terence D Murphy, Kim D Pruitt. 2016. "Reference Sequence (RefSeq) Database at NCBI: Current Status, Taxonomic Expansion, and Functional Annotation." *Nucleic Acids Research* 44 (D1): D733-45. https://doi.org/10.1093/nar/gkv1189.

Ozata, Deniz M, Ildar Gainetdinov, Ansgar Zoch, Dónal O'Carroll, and Phillip D Zamore. 2019. "PIWI-Interacting RNAs: Small RNAs with Big Functions." *Nature Reviews. Genetics* 20 (2): 89–108. https://doi.org/10.1038/s41576-018-0073-3.

Palazzo, Alexander F, and T Ryan Gregory. 2014. "The Case for Junk DNA." Edited by Joshua M. Akey. *PLoS Genetics* 10 (5): e1004351. https://doi.org/10.1371/journal.pgen.1004351.

Pan, Qun, Ofer Shai, Leo J Lee, Brendan J Frey, and Benjamin J Blencowe. 2008. "Deep Surveying of Alternative Splicing Complexity in the Human Transcriptome by High-Throughput Sequencing." *Nature Genetics* 40 (12): 1413–15. https://doi.org/10.1038/ng.259.

Poran, Asaf, Christopher Nötzel, Omar Aly, Nuria Mencia-Trinchant, Chantal T Harris, Monica L Guzman, Duane C Hassane, Olivier Elemento, and Björn FC Kafsack. 2017. "Single-Cell RNA Sequencing Reveals a Signature of Sexual Commitment in Malaria Parasites." *Nature* 551 (7678): 95–99. https://doi.org/10.1038/nature24280.

Rapsomaniki, Maria Anna, Xiao-Kang Lun, Stefan Woerner, Marco Laumanns, Bernd Bodenmiller, and María Rodríguez Martínez. 2018. "CellCycleTRACER Accounts for Cell Cycle and Volume in Mass Cytometry Data." *Nature Communications* 9 (1): 632. https://doi.org/10.1038/s41467-018-03005-5.

Reid, Adam J, Arthur M Talman, Hayley M Bennett, Ana R Gomes, Mandy J Sanders, Christopher JR Illingworth, Oliver Billker, Matthew Berriman, and Mara Kn Lawniczak. 2018. "Single-Cell RNA-Seq Reveals Hidden Transcriptional Variation in Malaria Parasites." *ELife* 7. https://doi.org/10.7554/eLife.33105.

Runne, Heike, Alexandre Kuhn, Edward J Wild, Wirahpati Pratyaksha, Mark Kristiansen, Jeremy D Isaacs, Etienne Régulier, Mauro Delorenzi, Sarah J Tabrizi, and Ruth Luthi-Carter. 2007. "Analysis of Potential Transcriptomic Biomarkers for Huntington's Disease in Peripheral Blood." *Proceedings of the National Academy of Sciences of the United States of America* 104 (36): 14424–29. https://doi.org/10.1073/pnas.0703652104.

Saliba, Antoine-Emmanuel, Lei Li, Alexander J Westermann, Silke Appenzeller, Daphne A C Stapels, Leon N Schulte, Sophie Helaine, and Jörg Vogel. 2016. "Single-Cell RNA-Seq Ties Macrophage Polarization to Growth Rate of Intracellular Salmonella." *Nature Microbiology* 2 (November): 16206. https://doi.org/10.1038/nmicrobiol.2016.206.

Santoro, Silvia, Ignazio Diego Lopez, Raffaella Lombardi, Andrea Zauli, Ana Maria Osiceanu, Melissa Sorosina, Ferdinando Clarelli, Silvia Peroni, Daniele Cazzato, Margherita Marchi, Angelo Quattrini, Giancarlo Comi, Raffaele Adolfo Calogero, Giuseppe Lauria and Filippo Martinelli Boneschi. 2018. "Laser Capture Microdissection for Transcriptomic Profiles in Human Skin Biopsies." *BMC Molecular Biology* 19 (1): 7. https://doi.org/10.1186/s12867-018-0108-5.

Shadab, Mohammad, Sonali Das, Anindyajit Banerjee, Roma Sinha, Mohammad Asad, Mohd Kamran, Mithun Maji, Baijayanti Jha, Makaraju Deepthi, Manoharan Kumar, Abhishek Tripathi, Bipin Kumar, Saikat Chakrabarti and Nahid Ali. 2019. "RNA-Seq Revealed Expression of Many Novel Genes Associated With Leishmania Donovani Persistence and Clearance in the Host Macrophage." *Frontiers in Cellular and Infection Microbiology* 9: 17. https://doi.org/10.3389/fcimb.2019.00017.

Shalek, Alex K, Rahul Satija, Joe Shuga, John J Trombetta, Dave Gennert, Diana Lu, Peilin Chen, Rona S Gertner, Jellert T Gaublomme, Nir Yosef, Schraga Schwartz, Brian Fowler, Suzanne Weaver, Jing Wang, Xiaohui Wang, Ruihua Ding, Raktima Raychowdhury, Nir Friedman, Nir Hacohen, Hongkun Park, Andrew P May, Aviv Regev. 2014. "Single-Cell RNA-Seq Reveals Dynamic Paracrine Control of Cellular Variation." *Nature* 510 (7505): 363–69. https://doi.org/10.1038/nature13437.

Shih, Barbara B, Ajit J Nirmal, Denis J Headon, Arne N Akbar, Neil A Mabbott, and Tom C Freeman. 2017. "Derivation of Marker Gene Signatures from Human Skin and Their Use in the Interpretation of the Transcriptional Changes Associated with Dermatological Disorders." *The Journal of Pathology* 241 (5): 600–613. https://doi.org/10.1002/path.4864.

Siomi, Mikiko C, Kaoru Sato, Dubravka Pezic, and Alexei A Aravin. 2011. "PIWI-Interacting Small RNAs: The Vanguard of Genome Defence." *Nature Reviews. Molecular Cell Biology* 12 (4): 246–58. https://doi.org/10.1038/nrm3089.

Skelly, Daniel A, Galen T Squiers, Micheal A McLellan, Mohan T Bolisetty, Paul Robson, Nadia A Rosenthal, and Alexander R Pinto. 2018. "Single-Cell Transcriptional Profiling Reveals Cellular Diversity and Intercommunication in the Mouse Heart." *Cell Reports* 22 (3): 600–610. https://doi.org/10.1016/j.celrep.2017.12.072.

Speranza, Emily, Brandi N Williamson, Friederike Feldmann, Gail L Sturdevant, Lizzette Pérez-Pérez, Kimberly Meade-White, Brian J Smith, Jamie Lovaglio, Craig Martens, Vincent J Munster, Atsushi Okumura, Carl Shaia, Heinz Feldmann, Sonja M Best, Emmie de Wit. 2021. "Single-Cell RNA Sequencing Reveals SARS-CoV-2 Infection Dynamics in Lungs of African Green Monkeys." *Science Translational Medicine* 13 (578). https://doi.org/10.1126/scitranslmed.abe8146.

Ståhl, Patrik L, Fredrik Salmén, Sanja Vickovic, Anna Lundmark, José Fernández Navarro, Jens Magnusson, Stefania Giacomello, Michaela Asp, Jakub O Westholm, Mikael Huss, Annelie Mollbrink, Sten Linnarsson, Simone Codeluppi, Åke Borg, Fredrik Pontén, Paul Igor Costea, Pelin Sahlén, Jan Mulder, Olaf Bergmann, Joakim Lundeberg, Jonas Frisén. 2016. "Visualization and Analysis of Gene Expression in Tissue Sections by Spatial Transcriptomics." *Science (New York, N.Y.)* 353 (6294): 78–82. https://doi.org/10.1126/science.aaf2403.

Stallaert, Wayne, Yannick Brüggemann, Ola Sabet, Lisa Baak, Marina Gattiglio, and Philippe IH Bastiaens. 2018. "Contact Inhibitory Eph Signaling Suppresses EGF-Promoted Cell Migration by Decoupling EGFR Activity from Vesicular Recycling." *Science Signaling* 11 (541). https://doi.org/10.1126/scisignal.aat0114.

Stapels, Daphne AC, Peter WS Hill, Alexander J Westermann, Robert A Fisher, Teresa L Thurston, Antoine-Emmanuel Saliba, Isabelle Blommestein, Jörg Vogel, and Sophie Helaine. 2018. "Salmonella Persisters Undermine Host Immune Defenses during Antibiotic Treatment." *Science (New York, N.Y.)* 362 (6419): 1156–60. https://doi.org/10.1126/science.aat7148.

Tay, Yvonne, John Rinn, and Pier Paolo Pandolfi. 2014. "The Multilayered Complexity of CeRNA Crosstalk and Competition." *Nature* 505 (7483): 344–52. https://doi.org/10.1038/nature12986.

Tomczak, Katarzyna, Patrycja Czerwińska, and Maciej Wiznerowicz. 2015. "The Cancer Genome Atlas (TCGA): An Immeasurable Source of Knowledge." *Contemporary Oncology (Poznan, Poland)* 19 (1A): A68-77. https://doi.org/10.5114/wo.2014.47136.

Turro, Ernest, Shu-Yi Su, Ângela Gonçalves, Lachlan J M Coin, Sylvia Richardson, and Alex Lewin. 2011. "Haplotype and Isoform Specific Expression Estimation Using Multi-Mapping RNA-Seq Reads." *Genome Biology* 12 (2): R13. https://doi.org/10.1186/gb-2011-12-2-r13.

Valencia-Sanchez, Marco Antonio, Jidong Liu, Gregory J Hannon, and Roy Parker. 2006. "Control of Translation and MRNA Degradation by MiRNAs and SiRNAs." *Genes & Development* 20 (5): 515–24. https://doi.org/10.1101/gad.1399806.

Wang, Xi, and Murray J Cairns. 2014. "Understanding Complex Transcriptome Dynamics in Schizophrenia and Other Neurological Diseases Using RNA Sequencing." *International Review of Neurobiology* 116: 127–52. https://doi.org/10.1016/B978-0-12-801105-8.00006-0.

Wang, Zhong, Mark Gerstein, and Michael Snyder. 2009. "RNA-Seq: A Revolutionary Tool for Transcriptomics." *Nature Reviews. Genetics* 10 (1): 57–63. https://doi.org/10.1038/nrg2484.

Weber, Christian, Mikael Koutero, Marie-Agnes Dillies, Hugo Varet, Cesar Lopez-Camarillo, Jean Yves Coppée, Chung-Chau Hon, and Nancy Guillén. 2016. "Extensive Transcriptome Analysis Correlates the Plasticity of Entamoeba Histolytica Pathogenesis to Rapid Phenotype Changes Depending on the Environment." *Scientific Reports* 6: 35852. https://doi.org/10.1038/srep35852.

Wellen, Kathryn E, and Craig B Thompson. 2010. "Cellular Metabolic Stress: Considering How Cells Respond to Nutrient Excess." *Molecular Cell* 40 (2): 323–32. https://doi.org/10.1016/j.molcel.2010.10.004.

Westermann, Alexander J, Lars Barquist, and Jörg Vogel. 2017. "Resolving Host-Pathogen Interactions by Dual RNA-Seq." *PLoS Pathogens* 13 (2): e1006033. https://doi.org/10.1371/journal.ppat.1006033.

Westermann, Alexander J, Konrad U Förstner, Fabian Amman, Lars Barquist, Yanjie Chao, Leon N Schulte, Lydia Müller, Richard Reinhardt, Peter F Stadler, and Jörg Vogel. 2016. "Dual RNA-Seq Unveils Noncoding RNA Functions in Host-Pathogen Interactions." *Nature* 529 (7587): 496–501. https://doi.org/10.1038/nature16547.

Wu, Hsing-Ju, Andrew HJ Wang, and Michael P Jennings. 2008. "Discovery of Virulence Factors of Pathogenic Bacteria." *Current Opinion in Chemical Biology* 12 (1): 93–101. https://doi.org/10.1016/j.cbpa.2008.01.023.

Xiao, Zhengyun, Guo Cheng, Yang Jiao, Chen Pan, Ran Li, Danmei Jia, Jing Zhu, Chao Wu, Min Zheng, and Junling Jia. 2018. "Holo-Seq: Single-Cell Sequencing of Holo-Transcriptome." *Genome Biology* 19 (1): 163. https://doi.org/10.1186/s13059-018-1553-7.

Yimthin, Thatcha, Jacqueline Margaret Cliff, Rungnapa Phunpang, Peeraya Ekchariyawat, Taniya Kaewarpai, Ji-Sook Lee, Clare Eckold, Megan Andrada, Ekkachai Thiansukhon, Kittisak Tanwisaid, Somchai Chuananont, Chumpol Morakot, Narongchai Sangsa, Wirayut Silakun, Sunee Chayangsu, Noppol Buasi, Nicholas Day, Ganjana Lertmemongkolchai, Wasun Chantratita, T. Eoin West and Narisara Chantratita. 2021. "Blood Transcriptomics to Characterize Key Biological Pathways and Identify Biomarkers for Predicting Mortality in Melioidosis." *Emerging Microbes & Infections* 10 (1): 8–18. https://doi.org/10.1080/22221751.2020.1858176.

Yoon, Je-Hyun, Kotb Abdelmohsen, and Myriam Gorospe. 2014. "Functional Interactions among MicroRNAs and Long Noncoding RNAs." *Seminars in Cell & Developmental Biology* 34 (October): 9–14. https://doi.org/10.1016/j.semcdb.2014.05.015.

Zimmermann, Michael, Maria Kogadeeva, Martin Gengenbacher, Gayle McEwen, Hans-Joachim Mollenkopf, Nicola Zamboni, Stefan Hugo Ernst Kaufmann, and Uwe Sauer. n.d. "Integration of Metabolomics and Transcriptomics Reveals a Complex Diet of Mycobacterium Tuberculosis during Early Macrophage Infection." *MSystems* 2 (4). https://doi.org/10.1128/mSystems.00057-17.

Part II. Understanding Cancer with Transcriptomics

Chapter 3

Transcriptomics in Basic Cancer Biology

Meenu A. Qureshi[1,*], Tawseeq Abdullah[1,2], Asiya Batool[1] and Nishawar Jan[1]

[1]Department of Biotechnology, Science Block, University of Kashmir, Hazratbal, Srinagar, Jammu and Kashmir, India
[2]Department of Biotechnology, Central University of Kashmir, Ganderbal, Jammu and Kashmir, India

Abstract

Despite the advancements made in medical science, cancer continues to present immense challenges worldwide as a dynamic, heterogeneous and lethal disease. With each passing day, awareness of its bio-molecular mechanisms, progresses and the path to cure is one of perseverance. Cancer is mechanistically driven by various genetic and epigenetic changes leading to the abnormal and disrupted expression of various critical genes. From time to time, extensive efforts have been made for the identification of these genetic alterations. However, various approaches in the form of cytogenetics, still-imaging, immuno-histological techniques and genomics for mutational screening have met with limited success. The limited knowledge hampers the proper understanding of these disease-causing mutations, culminating in the lack of agency in its detection, ultimately restricting proper clinical treatment. In response to these challenges, research efforts in the field of transcriptome biology have yielded a new hope in understanding various molecular abnormalities and alterations in different cancers.

Transcriptomics is a quantitative and qualitative branch of modern genetics that studies the total RNA transcripts in the cell, encompassing both coding and non-coding RNA using high throughput sequencing techniques. The qualitative gene expression profiling provides a more comprehensive understanding of genome, which allows a deeper insight into cancer mechanism by identifying molecular pathogenesis and

[*]Correspondence Author's Email: emmquee@gmail.com.

numerous peculiar epigenetic irregularities involved in various malignancies. Transcriptomics has enabled the detection of even subtle changes in genetic profile expressions, which helps to determine differences in types/subtypes within the same cancer, assisting in its characterization, classification and molecular profiling. This data can be incorporated into different tumor molecular profiles, leading to identification of druggable targets, besides providing additional therapeutic options to the patients with advanced cancers. Moreover, molecular classification of cancer based on genomic and transcriptomic alterations have led to the identification of molecules (cancer biomarkers) that are specific to tumors. These biomarkers carry huge clinical impact, with critical insights in the cancer research for accurate and earlier detection of cancers, positively impacting diagnosis, prognosis and improved surgical and therapeutic interventions. Transcriptomics based gene expression tests for prediction of tumor behaviour and drug response are emerging as methods of choice.

In this chapter, we discuss the advent of transcriptomics and its impact in cancer research and molecular therapy. We highlight its advantages over conventional cytogenetic, histological and DNA based hybridization techniques. We bring to light the recent technological advancements in transcriptomics and discuss the potential of transcriptomics in shaping the future of precision oncology. Finally, we describe the apparent challenges and limitations of this technique like cost-benefit analysis, adequacy of mRNA abundance, fitness of mRNA etc for accurate interpretation of the transcriptomic data.

Keywords: cancer biomarkers, drug targets, epigenetics, transcriptomics, tumour profiling

1. History of Gene Expression Techniques-Brief Introduction

Transcriptome refers to the total RNA content of an individual, while as transcriptomics is the study of an organism's transcriptome. Transcriptome profiling identifies genetic expression pattern of a cell that helps in understanding the genetic regulation of a particular cell type. In other words, it explains the biological state of a cell and divulges information about any dysregulation too. The techniques employed to study this content, termed as transcriptomics technologies have been widely used since late 1990s, though first attempts to study the transcriptome began earlier that decade. Transcripts were studied even before any transcriptomics approaches were in vogue. cDNA libraries were generated using reverse transcriptase from silk moth mRNAs in late 1970s (Sim et al. 1979, 1303–16). In the next decade of 1980s Sanger sequencing was used to sequence random transcripts from EST libraries. (Pan et al. 2008, 1413–15; Sutcliffe et al. 1982, 4942–46; Putney et al. 1983, 718–21). Later on, high throughput methods took over, like sequencing by synthesis (Solexa/Illumina, San Diego, CA). It was not till 1990s that Expressed Sequenced Tags (ESTs) became popular as a method of gene discovery as it offered the possibility of sequencing individual transcripts without sequencing the entire genome (Marra et al. 1998, 4–7).

Although techniques like Northern blotting, nylon membrane arrays, and reverse transcriptase quantitative PCR (RT-qPCR) (Alwine et al. 1977, 5350–54; Becker-Andre and Hahlbrock 1989, 9437–46) were used for the quantitative analysis of individual transcripts these techniques are considerably time and effort consuming and can only reveal information about a small section of a transcriptome (Corlan 2004). One of the sequence-based transcriptomic methods developed in 1995, based on Sanger's method was Serial Analysis of Gene Expression, in which transcripts were quantified by matching the fragments to a known gene. Further advancement in this technique created a SAGE variant using high throughput sequencing techniques, called digital gene expression analysis, was also used for some period (Wang et al. 2009, 57–63; Audic and Claverie 1997, 986–95).

Further, in mid 1990s and 2000s, some of the prominent and popular contemporary techniques microarray and RNA Seq were developed (Nelson 2001, 492–94; Wang et al. 2009, 57–63) Microarray is a technique that measures the abundance of a defined set of transcripts by their hybridisation to an array of complementary probes (Schena 1995, 467–70.); (Pozhitkov et al. 2007, 141–48). Transcriptomic analysis by microarray is a cost effective and labour-saving gene discovery method (Heller 2002, 129–53). With further advances in array manufacturing and designing, more and more genes were tested on a single array. Furthermore, the advent of fluorescence detection increased the sensitivity and accuracy for low abundance transcripts (McLachlan et al. 2005, 1-29); (Pozhitkov et al. 2007, 141–48). Hybridization-based approaches like microarray are high throughput and relatively inexpensive, however, they have several limitations viz, relying on previous knowledge of genomes, cross hybridisation causing high background levels (Okoniewski and Miller 2006, 1–14; Royce et al. 2007, e99) and the narrow range of detection due to signal saturation and background (Wang et al. 2009, 57–63).

Unlike microarray methods, sequence-based transcriptomic techniques directly determine the cDNA sequence. Initially, Sanger sequencing of cDNA or EST libraries (Boguski et al. 1994, 1993–94); (Gerhard et al. 2004, 2121–27) was used, which is an expensive and low throughput approach for gene analysis. Afterwards, tag-based sequencing methods were devised to overcome these shortcomings, which include serial analysis of gene expression SAGE; (Velculescu et al. 1995, 484–87) (Harbers and Carninci 2005, 495–502), cap analysis of gene expression (CAGE) (Kodzius et al. 2006, 211–22; Nakamura and Carninci 2004, 2688–93; Shiraki et al. 2003, 15776–81), and massively parallel signature sequencing (MPSS) (Brenner et al. 2000, 630–34; Peiffer et al. 2008, 1–16; Reinartz et al. 2002, 95–104). Although these approaches are high throughput furnishing 'digital' gene expression levels, they are on a higher side cost wise since Sanger sequencing is an expensive method. In addition, the sequence tags generated in SAGE/CAGE/MPSS fall short of being uniquely mapped to the reference genome and only a portion of transcript is analysed. Consequent upon these shortfalls, these traditional sequencing technologies do not provide a comprehensive transcriptome profile and gene annotation (Wang et al. 2009, 57–63). With the advent of next-generation high throughput sequencing technologies rapid and enormous advances have been made in the transcriptomics field, where RNA Seq has supplanting microarrays as the method of choice for transcriptional profiling (McGettigan 2013, 4–11).

2. RNA-Based Therapeutics

As a consequence of arrival of next-generation sequencing technologies rapid and massive advances have been made in the field of transcriptomics, where RNA Seq technology has gained popularity as the method of choice for transcriptional profiling (McGettigan 2013, 4–11). Among all the RNA based therapies major progress has been made incase of microRNAs (miRNAs). miRNAs are small non-coding RNAs that are involved in gene regulation at the post-transcriptional level, and function by interfering with the translation of target mRNAs. Thus, functionally, miRNAs can either be oncogenes (Onco miRNAs) or tumor suppressors (TS miRNAs). The dysregulation of miRNAs contributes to the pathogenesis of all types of cancer. In case of tumor suppressor miRNAs, for instance members of the Let-7 and miR-34 family, the diminished expression has been observed to promotes the tumor progression, invasion and metastasis (Hosseinahli et al. 2018, 5574–88).

Onco miRNAs (oncomiRs) (e.g., miR155 and miR21) are implicated in the promotion of cellular transformation, carcinogenesis, and metastasis, wherein they block the expression of tumor suppressor genes, and hence are valuable therapeutic targets in cancer treatment (Hemmatzadeh et al. 2016, 129–39) (Garzon et al. 2010, 775–89; Nedaeinia et al. 2017, 4129–40). On the other hand, research based evidences have reported the down regulation of TS miRNA in cancers (Osaki et al. 2011, 1123–30; Wang et al. 2019, 1–15). As a result, miRNAs are considered to be cancer diagnostic and prognostic markers, as well as effective therapeutic targets (Hosseinahli et al. 2018, 5574–88).

The miRNA activities can be regulated via two approaches inhibiting the oncogenic miRNA's expression and reviving the TS miRNA's expression (Hosseinahli et al. 2018, 5574–88) (Mollaei et al. 2019, 12369–84). In the former approach, referred to as miRNA inhibition therapy, antisense oligonucleotides with 2′-O-methyl groups (Weiler et al. 2006, 496–502); (Stenvang et al. 2012, 1–17); (Yu and Chen 2010, 137–48), 2′-O-methoxyethyl groups (Lima et al. 2018, 338–52) and other linkage blocking exonuclease degradation, such as *N*, *N*-diethyl-4-(4-nitronaphthalen-1-ylazo)-phenylamine (Lennox et al. 2013, e117) can be designed for inhibiting the expression of oncomiRs. Another strategy to suppress the oncomiR's expression relies on the use of locked nucleic acid (LNA) with sequences complementary to target miRNAs (Nedaeinia et al. 2017, 4129–40), (Chakraborty et al. 2016, 469–77). Presently, efficacy of LNA inhibitor of miR-155 (antimiR®) is undergoing clinical trials for treatment of chronic lymphocytic leukemia (CLL), adult T-cell lymphoma (ATLL), cutaneous T-cell lymphoma (CTCL) and diffuse large B-cell lymphoma (DLBCL) (*PRISM; SOLAR*; *Seto AG* et al, 2018, 428-444).

The later approach, referred to as miRNA replacement therapy, focuses on the restoration of TS miRNAs' expression (Bader et al. 2010, 7027–30). The therapy involves using either viral or nonviral miRNA delivery systems (Yang 2015, 179). The viral vectors are reported to be insufficient in evoking immune response (Nayak and Herzog 2010, 295–304). The non-viral vectors comprise of: polymer-based miRNA delivery systems (such as polyamidoamine dendrimers—PAMAMs (O'Neill 2020, 521; Ganju et al. 2017, 424–32); inorganic miRNA platforms (e.g., Fe_3O_4-based nanoparticles (Schade et al. 2013, 10710–26), gold or nanodiamonds platforms (Cao et al. 2013, 12120–25) and silica-based systems (Tivnan et al. 2012, e38129) or atelocollagen (Hao et al. 2016, 864–71); (Ishihara et al. 2016, 1-13) and lipid-based platforms (Wang et al. 2018, 1031–38.; Vickers and Remaley 2012, 91).

Since the discovery of miRNA in mid 1990, miRNAs have been extensively researched and thereby understood deeply. This understanding of the involvement of RNA in various critical cellular processes is imperative for the introduction of clinical, RNA-based techniques (Bonneau et al. 2019, 114). Also, a lot of research is on to overcome RNA therapeutic limitations such as, poor cellular uptake, rapid elimination or low bioavailability, and an augmented immune response (Baumann and Winkler 2014, 1967–84; Hanna et al. 2019, 478). Nonetheless, targeting RNA networks in cancer, has a great clinical potential. As a result, RNA based therapeutics are expected to grow rapidly in the coming years.

2.1. Tumor Heterogeneity and RNA-seq

Cancer is a complex and heterogeneous disease. The build-up of genetic mutations in the malignant cells and the complex interactions of cancer cells and other types of non- cancerous cells within the tumour microenvironment (TME) leads to the intra-tumoural heterogeneity which in turn determines disease progression and treatment sensitivity among patients (Wu et al. 2021, 1-11), (Fisher et al. 2013, 479–85), (Davis-Marcisak et al. 2019, 5102–12). Heterogeneity confers genetic and epigenetic variations, thus harbouring distinct molecular signatures with distinct biological behaviours and functions to the tumour cells. These genetic alterations can occur both between the different tumors (inter-tumoral heterogeneity) or within tumors of same histopathological subtype (intra-tumoral heterogeneity) (Fisher et al. 2013, 479–85). Tumour heterogeneity poses serious challenges for proper clinical diagnosis and treatment of cancers, and plays a significant role in resistance to anticancer therapies and disease recurrence. Intra-tumoral heterogeneity has implications for cancer treatment and discovery of biomarkers and carries huge potential for targeted treatments. The expansive understanding of TME and accurate characterization of genetic divergence and molecular variation which leads to heterogeneity in tumors is therefore critical for proper cancer diagnosis and advanced treatment which can further help to improve patient care and clinical outcomes (Davis-Marcisak et al. 2019, 5102–12).

The advancement in deep RNA sequencing, especially single cell RNA sequencing (scRNA-seq) is helpful in comprehensive gene expression profiling. Thus, unravelling the details regarding tumor ecosystem of an individual cell, such as the mechanism of tumor heterogeneity. Wu et al. (2021, 1-11) presented the comprehensive landscape of tumor ecosystem: cancer cells, immune cells, and stromal cells by profiling biopsies in patients with advanced NSC lung carcinoma using scRNA-seq. scRNA-seq combined with multi-sector biopsies were used to analyse the single cell expression profiles of individual glioma cells of different patients with brain tumors (Yu et al. 2020, 1306–18). Similarly single cell sequencing analysis was used to characterize the cell types inside the Renal Cell Carcinoma (RCC) (Zhang et al. 2021).

2.2. Transcriptomics in Clinical Classifications of Cancer

Transcriptomics technologies have increased our understanding of cancer at the molecular level. Whereas gene expression analysis assists researchers in investigating the molecular basis of cancer in a variety of ways, hence our knowledge of cancer progression is profoundly enhanced. Transcriptome analysis has contributed significantly in improving clinical classification of many tumors (Salem et al. 2017, 124–34; Golub et al. 1999, 531–37), in addition to elucidation of biochemical pathways, prediction of drug sensitivity and selective therapeutics (Clarke et al. 2004, 2560–91). Several gene expression studies have been carried out, to identify diagnostic and therapeutic targets with the aim of establishing molecular classification in breast cancer (Gruvberger et al. 2001, 5979–84; Stebbing and Wasan 2019, 1847-1850; Sandhu et al. 2010, 364–72; Reis-Filho and Pusztai 2011, 1812–23). A study by Perou et al. (2000, 747–52) was successful in characterisation of breast cancer, the results of which identified unique patterns of gene expression in each of the 65 samples analysed. Another study on breast cancer by Sørlie et al. (2001, 10869–74) where luminal type of breast cancer was further divided into two subtypes: luminal A and luminal B, depending upon the gene expression levels of markers like oestrogen receptor ER and Ki67 (Kondov et al. 2018, 961; Loi et al. 2007, 1239). Both the studies mentioned were carried out by using microarray technology, leading to the molecular classification of breast cancer currently in vogue, clinically (Chen et al. 2020, 1209–26).

Guinney et al. (2015, 1350–56) classified colorectal cancer (CRC) into four consensus molecular subtypes (CMS). Also, many clinical trials followed this particular classification and confirmed different treatment sensitivities, recurrence parameters and metastatic potential of individual CMS subtypes (Marisa et al. 2017, 3509; Okita et al. 2018, 18698; Lenz et al. 2017, 3511; Stintzing et al. 2017, 3510; Mooi et al. 2018, 2240–46). Such approaches are helpful in understanding the disease etiology and its management including diagnostic, prognostic and therapeutic strategies, also processes like drug development and clinical trial designing are further augmented (Stebbing and Wassan 2019, 1847-1850; Fontana et al, 2019, 520-527).

In case of high-grade serous ovarian cancer (HGSOC), the molecular profiling using RNA Seq revealed an overexpressed phenotype of PAX8, usually seen in cells within the distal fallopian tube (Xiang and Kong 2013, 735–38; Rodgers et al. 2016, 32785), which is in contrast to the classical view that it emerges from ovarian epithelial cells (Peres et al. 2018, 53–60); (Labidi-Galy et al. 2017, 1–11); (Wang et al. 2019, 194–201) Thus, RNA Seq analysis results, such as these have the potential to accelerate diagnostics and improve cancer management. Furthermore, cancer classification based on gene expression profile of various other cancers have also been made, viz uveal melanoma (Tschentscher et al. 2003, 2578–84), head and neck squamous cell carcinoma (Zivicova et al. 2018, 1279–90), large cell lung cancer (Karlsson et al. 2017, 1257–67.), and small round-blue cells tumors (neuroblastoma, rhabdomyosarcoma, non-Hodgkin lymphoma, Ewing tumor) (Khan et al. 2001, 673–79).

2.3. Identification of Cancer Biomarkers for Early Detection

Early diagnostic as well as treatment monitoring biomarkers are key components in cancer management. Carbohydrate antigen 125 (CA125), whose reliability as an ovarian cancer biomarker remains controversial (Elias et al. 2018, 903–14), has its specificity range between 35% and 91% (Pepin et al. 2014) with 50% raise in ovarian cancer (early onset cases) and approximately 90% increase at advanced stage (Moss et al. 2005, 308–12). Mosig et al. (2012, 1–8) performed transcriptomic analysis of 22 ovarian cancer samples and observed a considerable elevation in the levels of Insulin-like growth factor binding protein-4 (IGFBP-4) transcripts in early and later stages of the disease and also in relapsed samples, irrespective of CA125 levels.

In case of gastrointestinal cancers where the prognosis is poor often due to late diagnosis (Nassri et al. 2018), the discovery of relevant and sensitive biomarkers is imperative. The extremely specific protein biomarkers for esophageal squamous cell carcinoma (ESCC) which include squamous cell carcinoma antigen (SCC-Ag), carbohydrate antigen 19–9 (CA19–9), cytokeratin 19 fragment antigen 21–1 (Cyfra21–1), carcinoembryonic antigen (CEA) and carbohydrate antigen 72–4 (CA72–4) (Yang et al. 2019, 1-10), are marred by their decreased sensitivity, ranging only between 10% and 40%

(Zheng et al, 2014, 1-10). However, recent transcriptome studies have identified genes with increased expression that act as biomarkers for early detection. Xing et al. (2017, 2346) evaluated three serum markers for their clinical usage, out of more than 4000 differentially expressed genes in the ESCC transcriptome database, for their clinically usage: MMP13 (matrix metallopeptidase 13), SPP1 (secreted phosphoprotein 1) and CHI3L1(chitinase 3 like 1). The diagnostic panel comprising of MMP13, CHI3L1 and SPP1 identified around 90% of cases with early-onset ESCC, whileas the panel's detection level in case of non-cancerous tissues was between 10% and 15% (Xing et al. 2017, 2346).

Many RNAs have proven useful in precision oncology as they serve as independent cancer biomarkers (Xi et al. 2017, 9), e.g., transcriptional analysis of non-small-cell lung carcinoma (NSCLC) revealed that the two molecules up-regulated in NSCLC tumors: snoRNAs (Liao et al. 2010, 1–10) and 4-miRNA [miR-141 (microRNA 141), miR-200b (microRNA 200b), miR-193b (microRNA 193b) and microRNA 301 (miR-301) (Nadal et al. 2015, 1–9), can serve as early diagnostic markers of non-small-cell lung carcinoma (NSCLC).

In case of gastric cancer, microRNAs: miR-221, miR-20a and miR-106b were established as early detection biomarkers (Cai et al. 2013, 1-7; Toiyama et al. 2014, 43–57). In addition, long non-coding RNAs (lncRNAs), such as XIST, are believed to have possible applications in early detection of gastric cancer (Lu et al. 2017, 9545–52). Whileas numerous studies have also speculated the potential use of ncRNAs (noncoding RNAs) in CRC as well (Vychytilova-Faltejskova et al. 2018, 1019–28; Slaby 2016, 153–70). Another RNA molecule, the piRNA (Piwi-interacting RNA) that has role in transcriptional and post-transcriptional gene regulation, can serve as a biomarker in early diagnosis, treatment and prognosis in cancers like renal cancer (Busch et al. 2015, 1–11; Li et al. 2015, 381–88), hepatocellular carcinoma (Rizzo et al. 2016, 54650) and gastric cancer (Liu et al. 2019, 1–15; Liu et al. 2011, 784–91; Cui et al. 2011, 1050–57).

3. Applications of RNA-Sequencing in Cancer Research

Due to the popularity of high-throughput sequencing technologies, genomic data such as RNA-seq has become readily available. This has made significant contributions to a variety of fields, especially cancer science (Figure 1). Some of them are discussed below:

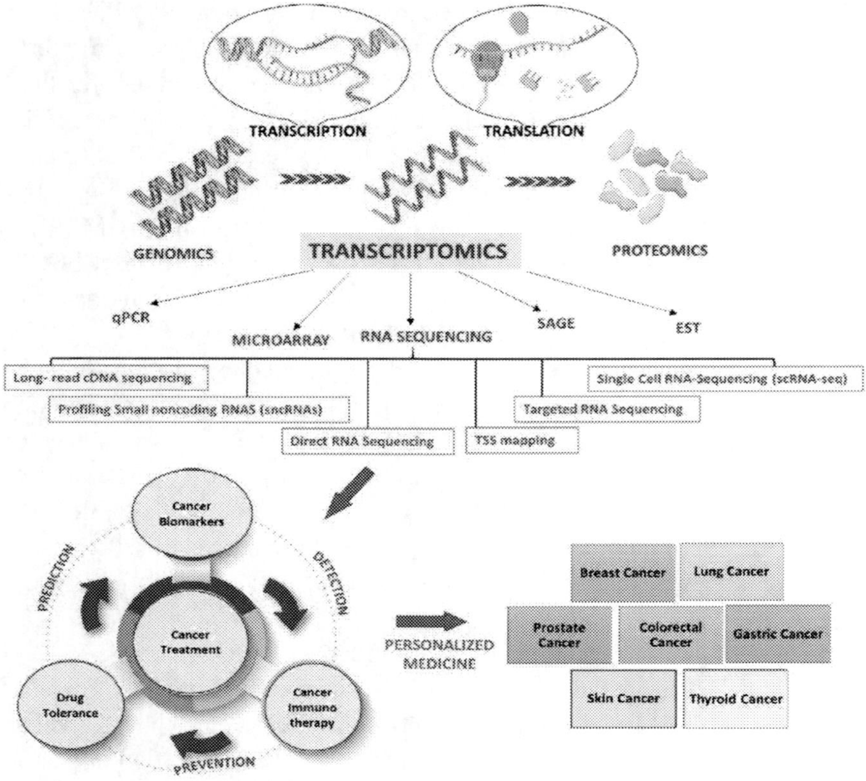

Figure 1. Role of transcriptomics in Cancer.

3.1. Cancer Variance and Adaptation

The transition of normal cells to cancer cells is always accompanied by heterogeneity, which over time may represent the progression and adaptation of cancer (Guo et al. 2019, 1–19.). Earlier RNA sequencing ignored variations in individual cells and observed all RNA transcripts in a given tissue or cell community. This difficulty is solved by transcriptome profiling of single-cell RNA sequencing, which provides single-cell resolution of the transcriptome. A recent single-cell RNA-seq analysis of 49 metastatic lung cancer samples showed improvements in plasticity caused by non-small cell lung cancer therapy, pointing to potential therapeutic options (Maynard et al. 2020, 1232–51). In melanoma, 4645 tumor cells from 19 patients were analyzed using single-cell RNA-seq, which included endothelial cells, mesenchymal cells, cancer cells, and immune cells. In contrast to scRNA-seq, another new technique called spatial transcriptome sequencing takes into account the cell position. The transcriptome of approximately 6750 tissue regions was studied in prostate cancer employing this technique, showing the whole-tissue transcriptional heterogeneity of whole multifocal prostate cancer and correctly defining the spectrum of cancer locus (Berglund et al. 2018). Since single-cell RNA sequencing can't be used on frozen samples and can't get all cell types in a tissue, single-nucleus RNA sequencing (snRNA-seq) has gotten a lot of coverage in recent years (Bakken et al. 2018, e0209648; Selewa et al. 2020, 1–13).

3.2. Cancer Drug Tolerance

One of the most common causes of cancer treatment relapse is drug tolerance. RNA sequencing has been a powerful tool in the discovery of cancer drug resistance mechanisms. Single-cell transcriptomics may be employed to investigate various chemoresistance mechanisms in cancer cells. It has revealed different mechanisms of trans-differentiation for homozygous and heterozygous phenotypic cells under drug selection (Sharma et al. 2018, 1–17). RNA Seq has now been used to do extensive studies on the drug susceptibility of haematological malignancies. Non-coding RNAs and fusion genes have been discovered to play a significant role in mediating drug resistance in haematological malignancies using RNA-Seq (Wang et al. 2019, 1–15). In neuroblastoma cells, RNA Seq has also shown that cancer cells express a high level of BORIS (brother of the regulator of imprinted sites), which controls the expression of survival genes during drug-resistance. Identification of such molecules may aid in our understanding of chemotherapy resistance and disease recurrence.

3.3. Differential Gene Expression and Cancer Biomarkers

Amongst the most common applications of RNA sequencing is differential gene expression analysis (Zubovic et al. 2020, 1–5). RNA sequencing can be used on samples from various contexts (different organisms, tissues, and time periods) to recognize differentially expressed genes, identifying their role and possible signalling targets (Robinson and Oshlack 2010, 1–9). Furthermore, differential gene expression assessment aids in the identification of new cancer biomarkers (Govindarajan et al. 2020, 1–20). Multiple reports suggest that gene fusions are closely linked to cancer progression and are valued as suitable cancer biomarkers (Wu et al. 2019, 385–90). RNA-CaptureSeq is the most commonly used method for detecting gene fusions in clinical samples as it has greater sequencing depth than whole transcriptome sequencing (Mercer et al. 2014, 989–1009). Gene fusions have the potential to stimulate the progression of cancer and provide specific targets for targeted therapy (Kloosterman et al. 2017, 3814–22).

Besides that, RNA Seq has led to the discovery of several special genomic factors, such as miRNA, lncRNA, and circRNA, that are commonly present in different forms of cancer (Sun and Chen 2020, 1–27); (Zhou et al. 2020, 1–15); (Liu et al. 2019, 1-15); (Liu et al. 2016, 20584). RNA-seq in coupled with western blotting and immunohistochemistry has identified some proteins as cancer biomarkers, like HER2 and COX2 (cyclooxygenase2), which can function as possible biomarkers for diagnosing colorectal cancer (Zhou et al. 2018, 301). RNA Seq can be used to investigate the application value of TMB (Tumor mutation burden) in diffuse glioma (Wang et al. 2020, 1–12). Moreover, copy number variations (CNV) can be inferred from scRNA-seq results, which can be used to differentiate malignant cells from non-malignant ones.

3.4. Cancer Immunotherapy

The cancer immune microenvironment is made up of Tumor Infiltrating Lymphocytes and their associations with stromal and malignant cells. Transcriptomic profiling, specifically scRNA-seq, offers detailed knowledge on cellular function and interactions in the tumor microenvironment (TME). scRNA-seq allows for molecular and genomic characterization of human immune cells both qualitatively and quantitatively to evaluate the cellular heterogeneity in the cancer microenvironment (Shalek et al. 2013, 236-40; Pan et al. 2019, 1–18; Macosko et al. 2015, 1202–14). scRNA-seq has led to the discovery of various immunomodulatory genes that inhibit T cell activity in AML patients (van Galen et al. 2019, 1265-81). It has also provided new insight into anti-cancer immunotherapy by targeting neutrophils (Janiszewska et al. 2019, 879–88). Besides this, intracellular staining and sequencing (INs-seq), a novel technology that combines intracellular protein activity measurements and scRNA-seq has unravelled novel

Arg1 + Trem2 + regulatory myeloid (Mreg) for myeloid-targeted immunotherapy (Katzenelenbogen et al. 2020, 872–85).

4. Application of Transcriptomics in Cancer Diagnostic

Here we focus on the potential and current clinical applications of gene expression profiling by cancer type. The potential biomarkers discovered through transcriptome analysis in different cancers have been summarized in Table 1.

4.1. Breast Cancer

In order to extract molecular signatures that identify breast cancer cases with diverse clinical results and therapy outcomes, gene analysis has been most successful (Goodison et al. 2010, 855–62; Rouzier et al. 2005, 5678–85). A microarray technique was used in a 2002 study to identify a 70-gene signature that predicted metastasis incase of breast cancer patients with negative lymph-node diagnosis. The findings of this study were used to develop Agendia's MammaPrint technology, potentially able to produce a risk index score in breast cancer with both positive and negative hormone-receptor (Harbeck et al. 2014, 434–44). Analogous studies have established a 21-gene signature (Oncotype DX) that quantifies a woman's ductal carcinoma in situ (DCIS) recurrence risk as well as the probability of benefit from chemo and radiotherapy (Paik et al. 2006, 3726–34); Solin et al. 2013, 701–10). Mammaprint and Oncotype have first entered clinical use in 2006 and 2007, respectively, with Oncotype DX being the most commonly used breast cancer assay (78 percent of respondents using it) in Europe, and 34% using the Mammaprint. (Mulcahy, 2014). CTCs (circulating tumor cells) have recently been identified as a possible source of prognostic markers for diagnosing and managing metastatic breast cancer patients (Wang et al. 2013, 1031–38). Twelve genes have shown high expression in CTCs of metastatic bone patients relative to their healthy counterparts, with TFF1 (Trefoil factor 1) being the most differentially expressed gene in cases of bone metastases (Wang et al. 2013, 1031–38). Another prospective study has discovered 32 candidate genes, which on analysis appears to be linked to breast cancer clinical outcomes. A six-gene signature for disease recurrence (GABRP, ESR1,

TBC1D9, PGR, LRBA and SLC39A6) a five-gene signature for disease mortality (GABRP, PGR, LRBA, TBC1D9, and SLC39A6) and were also discovered. These studies point to the possibility that CTCs could help predict clinical breast cancer outcomes on a large scale (Andres et al. 2013, 1–18).

4.2. Colorectal Cancer

miRNAs have been seen to function either as oncogenes or tumor suppressors with potential prognostic and diagnostic value (Schetter et al. 2012, 244). Many miRNAs have shown dysregulation during colorectal cancer. Although just a handful has been functionally characterized, miR-375 appears to be crucial for regulating apoptosis and cellular growth in CRCs. A prospective analysis discovered that miR-375 expression inhibits tumor growth via its two predominant downstream targets-HELLS and NOLC1 in CRC (Christensen et al. 2014, e96767). Colorectal cancer patients share a metabolic signature with tremendous prognostic and therapeutic promise, considering their diverse genetic backgrounds, pathologic stages, defects, and geographic locations (Qiu et al. 2014, 2136–46). Agendia and Genomic Health detected an 18-gene and a 12-gene CRC biosignature respectively and transformed them into diagnostic tools. Coloprint, a genome microarray that utilizes 18 genes to determine relapse risk in stage II colon cancer patients, has lately been verified in a population of 135 patients (Maak et al. 2013, 1053–58). Another 12-gene Oncotype DX assay (Yothers et al. 2013, 4512.) employs RT-qPCR to provide prognostic information on an increased risk of disease recurrence, T stage, tumor grade, mismatch repair (MMR) evaluations, lymphovascular invasion, and the number of nodes tested (Yothers et al. 2013, 4512.). GeneFx Colon (Precision Therapeutics), OncoDefender-CRC (Everist Genomics, Inc) and ColonPRS (Signal Genetics LLC), are three other tests that have recently been launched (Kopetz and Abbruzzese 2009, 7451–52).

Table 1. Potential biomarkers discovered in different cancers using transcriptome analysis

Type of Cancer	Biomarkers	Expression in Cancer Tissue	*Ref*
Gastric cancer	miR-1, miR-20a, miR-27a, miR-34a, miR-423-5p, miR-106b, miR-20a, miR-21, miR-221, miR-451, miR-17-5p, miR-21, miR-106a, miR-106b	up	*Cai H et al, 2013, 1–7 Tsujiura M et al, 2010, 1174–1179*
	piR-651	down	*Cui L et al. 2011, 1050–57*
Breast cancer	miR-1, miR-20a, miR-27a, miR-34a, miR-423-5p, miR-17-5p, miR-21, miR-106a, miR-106b	up	*Kahraman M et al, 2018, 1-11*
Esophageal cancer	MMP13, CHI3L1, SPP1	up	*Xing et al., 2017, 2346*
Colorectal cancer	miR-17-92a, miR-135, miR-21	up	*Koga Y et al, 2010, 1435–42*
	piR-5937, piR-28876, piR-23210, piR-32159, piR-651	down	*Vychytilova-Faltejskova P et al, 2018, 1019–28.*
	ABCD3	down	*Yu T et al, 2018, 48-54*
Ovarian cancer	IGFBP-4	up	*Mosig RA et al, 2012, 1-8*
Gallbladder carcinoma	BIRC5	up	*Gu X et al, 2020, 475-84*
Glioblastoma multiforme	Thymosin β4 (TMSB4X), S100A10	up	*Cheng et al, 2020, 167*
	miR-20a, miR-106a	up	*Huang S.W et al, 2018, 1405-13*
Hogdkins lymphoma	piR-651	down	*Cordeiro et al, 2016, 1154*
Hepatocellular cancer	piR-013306	up	*Rizzo et al, 2016, 54650*
Multiple myeloma	piR-823	up	*Li et al, 2019, 5227-38*
Melanoma	Ro-associated Y RNA (YRNAs): RNY3P1, RNY4P1, RNY4P25 (upregulation) miR-320a-3p, miR-134-5p (downregulation)	up/down	*Solé et al, 2019, 70*
	miR-21	up	*Saldanha et al, 2013, 1381-84*
Renal cancer	PIWIL1, piR-823	down	*Iliev et al, 2016, 6419-23*
Non-small cell lung carcinoma	4-miRNA, snoRNA	up	*Nadal et al. 2015, 1-9*
	34 miRNA	up/down	*Bianchi F et al, 2011, 495-503*
Pancreatic cancer	KRAS mRNA (as salivary biomarker), LAMC2	up	*Kosanam H, et al, 2013, 2820-32*

4.3. Lung Cancer

In many tumor forms, including NSCLC, a gene signature of "stemness" originating from BMI-1-regulated genes in normal stem cells is linked to metastasis and survival (Glinsky 2008, 2846–53). STAT1, ERBB3, DUSP6, MMD, and LCK are other gene signatures linked to lung cancer (Toffalorio et al. 2014, 1244–49.). Centred on the expression of the genes GC and CEACAM1, highly sensitive and reproducible immunohistochemistry (IHC) assay with a new algorithm has been established to differentiate between normal and aberrant cells in lung carcinoma (Toffalorio et al. 2014, 1244–49). In addition, Myriad Genetics developed the myPlan Lung Cancer assay after identifying a distinct 46-gene biosignature as a measure of lung cancer prognosis. For stage I and II tumors, the myPlan RT-PCR-based assay produces a score that predicts the 5-year probability of lung adenocarcinoma death (Wistuba et al. 2013, 6261–71). HistoPlus, an RT-qPCR-based assay, is an available related tool that identifies a 57-gene biosignature to discriminate between squamous cell carcinoma, small cell carcinoma, adenocarcinoma and carcinoid (Wilkerson et al. 2013, 485–97).

4.4. Prostate Cancer

Transcriptome profiling has proven to be highly successful while studying prostate cancer cell lines, which was formerly challenging owing to the inadequate number of circulating tumor cells. Research-based study revealed that using commercially available RT-qPCR, gene profile can be accurately obtained from single prostate cancer cell lines derived from bone marrow (Welty et al. 2013, 1-11). RNA-Seq has been used in obtaining a systematic map of transcriptome profiles of 14 primary prostate cancers, revealing abnormal alternate splicing, repeated gene fusions, and cancer-associated lnRNAs (Ren et al. 2012, 806–21).

Also, the Oncotype DX prostate cancer assay was created with a 12-gene biosignature. The RT-qPCR assay results are used to measure a GPS (Genomic Prostate Score), that can be a potential prognostic predictor of prostate cancer (Knezevic et al. 2013, 1–12). The Prolaris assay, on the other hand, looks at the expression of 31 genes involved in cell cycle progression and could be a valuable predictor of prostate cancer-specific survival and biochemical recurrence (Cuzick et al. 2012, 1095–99).

4.5. Skin Cancer

Skin cancer is a severely complicated condition that encompasses a wide spectrum of SPSs (skin pathological states). Transcriptomic technologies like microarray and RNA-Seq (Lowe et al. 2017, e1005457) has opened up new avenues for skin cancer research. This advancement, in particular, may enhance knowledge of the possible biological links and distinctions among SPSs which can be identified in simultaneously in a simple analysis. As a result, needless drugs or medical procedures such as excision operations, radiation therapies, or pharmaceutical administration can be avoided. RNA-seq study of cutaneous squamous cell carcinoma (cSCC) and healthy skin has found a substantial number of differently expressed transcripts, including mRNAs, circRNAs and lncRNAs with previously unknown functions in cSCC (Das Mahapatra et al. 2020, 1-12). Also, a comprehensive transcriptome profiling analysis of human keratinocytes exposed to various UVR (ultraviolet radiation) settings has described the kinetic alterations in global gene expression. Based on UV-responsiveness and variations in their mRNA levels before and post exposure, the researchers observed a group of UV-responsive genes, referring to them as UV signature genes (UVSGs). Computational Gene Set Enrichment Analysis has proven that a substantial percentage of the UVSGs are exclusively dysregulated in human skin squamous cell carcinomas, an observation not found in other human malignancies. Thus, UVSGs can be potential as well as specific candidates for clinical diagnosis of skin cancer (Shen et al. 2016, e0163054).

4.6. Thyroid Cancer

Thyroid cancer is one of the most common endocrine system malignancies with differentiated thyroid cancer (DTC) being the most frequent type developing from thyroid follicular cells. The majority of thyroid malignancies are papillary and follicular thyroid cancer and most patients have a favourable prognosis. A slew of research has effectively identified the underlying genetic basis of human malignancies due to the continuous increase in next-generation sequencing (NGS) technology (Van Dijk et al. 2014, 418–26). Early breakthroughs in identifying and targeting specific oncogenic drivers, along with the growing practicality of sequencing tumor genomes, have ushered in the promise of genome-driven thyroid cancer treatment.

5. Challenges and Limitations

The processing of tissue samples and the use of modern computational approaches are the two key obstacles connected with transcriptomic analysis. In order to be optimally implemented in clinical practice, transcriptomic analysis requires significant optimization, validation, and cost reduction. Following are some of the challenges faced by transcriptomic analysis.

5.1. Construction of Library

RNA-Seq has only a few steps, but it does need numerous modification phases during the generation of cDNA libraries, that can make it difficult to employ in profiling all kinds of transcripts. RNA and cDNA fragmentation procedures (sonication, nebulation or RNA hydrolysis, DNase I treatment, etc.) produce distinct biases in the results. The interpretation of RNA-Seq data is additionally complicated by several alterations used during library construction. Many short reads that are similar to one other, for example, can be acquired from amplified cDNA libraries. These might be PCR artifacts or they might be a true reflection of numerous RNA species. Identifying if the same sequences appear in various biological replicates is one approach to distinguish between these options. Another important aspect in library creation is whether to develop strand-specific libraries, as two studies have done (Lister et al. 2008, 523–36; Cloonan et al. 2008, 613–19). Furthermore, it is critical to guarantee that the antisense transcripts are not reverse transcription artifacts (Wu et al. 2008, 1–14). For these issues, the majority of investigations to date have examined cDNAs without strand information.

5.2. Cost vs Coverage

Another crucial consideration is sequence coverage, or the proportion of transcripts examined, which has financial consequences. More sequencing depth is required for greater coverage. It takes a lot of data to find a rare transcript or variation. In principle, the bigger the genome and the more intricate the transcriptome, the greater the depth of sequencing required for appropriate coverage. Despite the fact that the cost and accuracy of genome sequencing have greatly improved, few people choose to have their genome sequenced. Although significant cost savings in relative terms, the process is thought to be costly. Molecular diagnostics used to be poorly covered by insurance, but improvements in billing processes are in the works that might alter that.

5.3. Bioinformatic Challenges

RNA-Seq, like most other high-throughput sequencing techniques, encounters a myriad of informatics hurdles, such as the implementation of effective methods for storing, retrieving, and processing large amounts of data. These challenges need to be addressed in order to minimize the error in base-calling and image analysis, as well as to eliminate low-quality reads. The initial step in data processing is to align the short reads from RNA-Seq to the target genome, or to compile them into contigs. However, reads that bridge exon junctions or include poly(A) ends are also included in the short transcriptome reads and cannot be analysed in the same way. The development of computationally easy algorithms to discover unique splicing events that occur between two different sequences or between exons from two separate genes will be a future challenge.

Conclusion and Future Prospects

On both the molecular and clinical levels, transcriptomics deserves to be regarded a modern "gamechanger" in cancer. A revolution in the knowledge, development, and therapeutic use of RNA-based methods has occurred in recent years. Transcriptomics has shown a variety of different transcriptional programs that exist inside each cell. The continual advancement of molecular approaches for transcriptome analysis and the adoption of NGS platforms have significantly altered the landscape of RNA research across the world. In clinical oncology, a number of predictive and prognostic gene expression tests, as well as cancer biomarkers, have been used to date. The shift in cancer categorization from histopathological to molecular allows for a more customized approach to tumor diagnosis and treatment. The development of transcriptome profiling will enable uniformity and price reduction in its evaluation, paving the way for transcriptomics for becoming a scripture of modern cancer treatment.

RNA-seq is playing an increasingly important role in cancer patient treatment, along with preclinical research and fundamental science. It has spawned a slew of transcriptome-based customized oncology studies. RNA-Seq has proved to be quite useful in identifying changes in transcript quantity and/or structure in cancer, but there are several difficulties and possibilities ahead. Comprehending tumor heterogeneity and complexity, and also the genome's dynamic expression, necessitates the use of numerous novel methods. To completely understand tumor immunological and -omic profiles, improve comprehensive tumor molecular profiles, and guide treatment decisions, precision medicine will require the integration of genomes, transcriptomics, proteomics, and epigenetics. Clinical studies can verify and strengthen the data available by looking at the role of transcriptomics in choosing suitable targeted therapies, asserting that the use of transcriptomics can expand the number of patients treated with matched targeted therapy and help propel to favourable outcomes, thus offering the next frontier for precision medicine.

References

Alwine, James C, David J Kemp, and George R Stark. 1977. "Method for Detection of Specific RNAs in Agarose Gels by Transfer to Diazobenzyloxymethyl-Paper and Hybridization with DNA Probes." *Proceedings of the National Academy of Sciences of the United States of America* 74 (12): 5350–54. https://doi.org/10.1073/pnas.74.12.5350.

Andres, Sarah A, Guy N Brock, and James L Wittliff. 2013. "Interrogating Differences in Expression of Targeted Gene Sets to Predict Breast Cancer Outcome." *BMC Cancer* 13 (1): 1–18. https://doi.org/10.1186/1471-2407-13-326.

Audic, Stéphane, and Jean Michel Claverie. 1997. "The Significance of Digital Gene Expression Profiles." *Genome Research* 7 (10): 986–95. https://doi.org/10.1101/gr.7.10.986.

Bader, Andreas G, David Brown, and Matthew Winkler. 2010. "The Promise of MicroRNA Replacement Therapy." *Cancer Research. AACR* 70 (18): 7027–30. https://doi.org/10.1158/0008-5472.CAN-10-2010.

Bakken, Trygve E, Rebecca D Hodge, Jeremy A Miller, Zizhen Yao, Thuc Nghi Nguyen, Brian Aevermann, Eliza Barkan, et al. 2018. "Single-Nucleus and Single-Cell Transcriptomes Compared in Matched Cortical Cell Types." *PLoS ONE* 13 (12): e0209648. https://doi.org/10.1371/journal.pone.0209648.

Baumann, Volker, and Johannes Winkler. 2014. "MiRNA-Based Therapies: Strategies and Delivery Platforms for Oligonucleotide and Non-Oligonucleotide Agents." *Future Medicinal Chemistry. Future Science.* 6(17):1967-84 https://doi.org/10.4155/fmc.14.116.

Becker-André, Michael, and Klaus Hahlbrock. 1989. "Absolute MRNA Quantification Using the Polymerase Chain Reaction (PCR). A Novel Approach by a PCR Aided Transcipt Titration Assay (PATTY)." *Nucleic Acids Research* 17 (22): 9437–46. https://doi.org/10.1093/nar/17.22.9437.

Berglund, Emelie, Jonas Maaskola, Niklas Schultz, Stefanie Friedrich, Maja Marklund, Joseph Bergenstråhle, Firas Tarish, et al. 2018. "Spatial Maps of Prostate Cancer Transcriptomes Reveal an Unexplored Landscape of Heterogeneity." *Nature Communications* 9 (1): 1–13. https://doi.org/10.1038/s41467-018-04724-5.

Bianchi, Fabrizio, Francesco Nicassio, Matteo Marzi, Elena Belloni, Valentina Dall'Olio, Loris Bernard, Giuseppe Pelosi, Patrick Maisonneuve, Giulia Veronesi, and Pier Paolo Di Fiore. 2011. "A serum circulating miRNA diagnostic test to identify asymptomatic high-risk individuals with early stage lung cancer.*" EMBO molecular medicine* 3 (8): 495-503. https://doi.org/10.1002/emmm.201100154.

Boguski, M. S., C. M. Tolstoshev, and D. E. Bossett. 1994. "Gene discovery in dbEST." *Science* (Wash. DC) 265 (5181): 1993-1994. doi: 10.1126/science.8091218.

Bonneau, E, B Neveu, E Kostantin, G J Tsongalis, and V De Guire. 2019. "How Close Are MiRNAs from Clinical Practice? A Perspective on the Diagnostic and Therapeutic Market." *Ejifcc* 30(2):114.

Brenner, Sydney, Maria Johnson, John Bridgham, George Golda, David H Lloyd, Davida Johnson, Shujun Luo, et al. 2000. "Gene Expression Analysis by Massively Parallel Signature Sequencing (MPSS) on Microbead Arrays." *Nature Biotechnology* 18 (6): 630–34. https://doi.org/10.1038/76469.

Busch, Jonas, Bernhard Ralla, Monika Jung, Zofia Wotschofsky, Elena Trujillo-Arribas, Philipp Schwabe, Ergin Kilic, Annika Fendler, and Klaus Jung. 2015. "Piwi-Interacting RNAs as Novel Prognostic Markers in Clear Cell Renal Cell Carcinomas." *Journal of Experimental and Clinical Cancer Research* 34 (1): 1–11. https://doi.org/10.1186/s13046-015-0180-3.

Cai, Hui, Yuan Yuan, Yun Fei Hao, Tian Kang Guo, Xue Wei, and Ying Mei Zhang. 2013. "Plasma MicroRNAs Serve as Novel Potential Biomarkers for Early Detection of Gastric Cancer." *Medical Oncology* 30 (1): 1–7. https://doi.org/10.1007/s12032-012-0452-0.

Cao, Minjun, Xiongwei Deng, Shishuai Su, Fang Zhang, Xiangqian Xiao, Qin Hu, Yongwei Fu, et al. 2013. "Protamine Sulfate-Nanodiamond Hybrid Nanoparticles as a Vector for MiR-203 Restoration in Esophageal Carcinoma Cells." *Nanoscale* 5 (24): 12120–25. https://doi.org/10.1039/c3nr04056a.

Chakraborty, Chiranjib, Zhi-Hong Wen, Govindasamy Agoramoorthy, and Chan-Shing Lin. 2016. "Therapeutic MicroRNA Delivery Strategies with Special Emphasis on Cancer Therapy and Tumorigenesis: Current Trends and Future Challenges." *Current Drug Metabolism* 17 (5): 469–77. https://doi.org/10.2174/1389200217666160126142408.

Chen, Teng, Julia Y. S. Tsang, Xiao Chun Su, Peng Li, Wen Qin Sun, Iris L. K. Wong, Kit Ying Choy, et al. 2020. "SALL4 Promotes Tumor Progression in Breast Cancer by Targeting EMT." *Molecular Carcinogenesis* 59 (10): 1209–26. https://doi.org/10.1002/mc.23250.

Cheng, Quan, Jing Li, Fan Fan, Hui Cao, Zi-Yu Dai, Ze-Yu Wang, and Song-Shan Feng. 2020. "Identification and analysis of glioblastoma biomarkers based on single cell sequencing." *Frontiers in bioengineering and biotechnology* 8: 167. https://doi.org/10.3389/fbioe.2020.00167.

Christensen, Lise Lotte, Anja Holm, Juha Rantala, Olli Kallioniemi, Mads H Rasmussen, Marie S Ostenfeld, Frederik Dagnaes-Hansen, et al. 2014. "Functional Screening Identifies MiRNAs Influencing Apoptosis and Proliferation in Colorectal Cancer." *PLoS ONE* 9 (6): e96767. https://doi.org/10.1371/journal.pone.0096767.

Clarke, Paul A, Robert Te Poele, and Paul Workman. 2004. "Gene Expression Microarray Technologies in the Development of New Therapeutic Agents." *European Journal of Cancer* 40 (17 SPEC. ISS.): 2560–91. https://doi.org/10.1016/j.ejca.2004.07.024.

Cloonan, Nicole, Alistair R. R. Forrest, Gabriel Kolle, Brooke B.A. Gardiner, Geoffrey J Faulkner, Mellissa K Brown, Darrin F Taylor, et al. 2008. "Stem Cell Transcriptome Profiling via Massive-Scale MRNA Sequencing." *Nature Methods* 5 (7): 613–19. https://doi.org/10.1038/nmeth.1223.

Cordeiro, Anna, Mariano Monzó, and Alfons Navarro.2017. "Non-coding RNAs in Hodgkin lymphoma." *International journal of molecular sciences* 18 (6) 1154. https://doi.org/10.3390/ijms18061154.

Corlan, Alexandru Dan. "Medline trend: automated yearly statistics of PubMed results for any query, 2004." *Web resource at URL: http://dan. corlan. net/medline-trend. html.* The full trend can be viewed here (http://dan.corlan.net/cgi-bin/medline-trend.

Cui, Long, Yanru Lou, Xinjun Zhang, Hui Zhou, Hongxia Deng, Haojun Song, Xiuchong Yu, Bingxiu Xiao, Weihua Wang, and Junming Guo. 2011. "Detection of Circulating Tumor Cells in Peripheral Blood from Patients with Gastric Cancer Using PiRNAs as Markers." *Clinical Biochemistry* 44 (13): 1050–57. https://doi.org/10.1016/j.clinbiochem.2011.06.004.

Cuzick, J, D M Berney, G Fisher, D Mesher, H Møller, J E Reid, M Perry, et al. 2012. "Prognostic Value of a Cell Cycle Progression Signature for Prostate Cancer Death in a Conservatively Managed Needle Biopsy Cohort." *British Journal of Cancer* 106 (6): 1095–99. https://doi.org/10.1038/bjc.2012.39.

Davis-Marcisak, Emily F, Thomas D Sherman, Pranay Orugunta, Genevieve L Stein-O'Brien, Sidharth V Puram, Evanthia T. Roussos Torres, Alexander C Hopkins, et al. 2019. "Differential Variation Analysis Enables Detection of Tumor Heterogeneity Using Single-Cell RNA-Sequencing Data." *Cancer Research* 79 (19): 5102–12. https://doi.org/10.1158/0008-5472.CAN-18-3882.

Fisher, Rosie, Lazos Pusztai, and C Swanton. 2013. "Cancer Heterogeneity: Implications for Targeted Therapeutics." *British Journal of Cancer.* 108: 479–85. https://doi.org/10.1038/bjc.2012.581.

Fontana, E., K. Eason, Andrés Cervantes, R. Salazar, and Anguraj Sadanandam. "Context matters—consensus molecular subtypes of colorectal cancer as biomarkers for clinical trials." *Annals of Oncology* 30, 4 (2019): 520-27. doi: 10.1093/annonc/mdz052.

Forstner, Rosemarie. 2020. "Early Detection of Ovarian Cancer." *European Radiology. Elsevier* 30: 5370–73. https://doi.org/10.1007/s00330-020-06937-z.

Galen, Peter van, Volker Hovestadt, Marc H. Wadsworth, Travis K Hughes, Gabriel K Griffin, Sofia Battaglia, Julia A Verga, et al. 2019. "Single-Cell RNA-Seq Reveals AML Hierarchies Relevant to Disease Progression and Immunity." *Cell* 176 (6): 1265-81.e24. https://doi.org/10.1016/j.cell.2019.01.031.

Ganju, Aditya, Sheema Khan, Bilal B Hafeez, Stephen W Behrman, Murali M Yallapu, Subhash C Chauhan, and Meena Jaggi. 2017. "MiRNA Nanotherapeutics for Cancer." *Drug Discovery Today. Elsevier* 22 (2) 424-32. https://doi.org/10.1016/j.drudis.2016.10.014.

Garzon, Ramiro, Guido Marcucci, and Carlo M Croce. 2010. "Targeting MicroRNAs in Cancer: Rationale, Strategies and Challenges." *Nature Reviews Drug Discovery. Nature Publishing Group*. 9: 775–89. https://doi.org/10.1038/nrd3179.

Gerhard, Daniela S, Lukas Wagner, Elise A Feingold, Carolyn M Shenmen, Lynette H Grouse, Greg Schuler, Steven L Klein, Susan Old, Rebekah Rasooly, and Peter Good. 2004. "The Status, Quality, and Expansion of the NIH Full-Length CDNA Project: The Mammalian Gene Collection (MGC)." *Genome Research* 14 (10 B): 2121–27. https://doi:10.1101/gr.2596504.

Glinsky, Gennadi V. 2008. "'Stemness' Genomics Law Governs Clinical Behavior of Human Cancer: Implications for Decision Making in Disease Management." *Journal of Clinical Oncology. American Society of Clinical Oncology* 26(17): 2846-53. https://doi.org/10.1200/JCO.2008.17.0266.

Goldstein, Darlene R. 2005. "Analyzing Microarray Gene Expression Data." *Journal of the American Statistical Association* 100 (472): 1464–65. https://doi.org/10.1198/jasa.2005.s60.

Golub, Todd R, Donna K Slonim, Pablo Tamayo, Christine Huard, Michelle Gaasenbeek, Jill P Mesirov, Hilary Coller, et al. 1999. "Molecular Classification of Cancer: Class Discovery and Class Prediction by Gene Expression Monitoring." *Science* 286 (5439): 531–37. https://doi.org/10.1126/science.286.5439.531.

Goodison, Steve, Yijun Sun, and Virginia Urquidi. 2010. "Derivation of Cancer Diagnostic and Prognostic Signatures from Gene Expression Data." *Bioanalysis. Future Science* 2(5):855-62. https://doi.org/10.4155/bio.10.35.

Govindarajan, Meinusha, Christoph Wohlmuth, Matthew Waas, Marcus Q Bernardini, and Thomas Kislinger. 2020. "High-Throughput Approaches for Precision Medicine in High-Grade Serous Ovarian Cancer." *Journal of Hematology and Oncology*. Springer 13(1):134. https://doi.org/10.1186/s13045-020-00971-6.

Gruvberger, Sofia, A. Borg, Mårten Fernö, Markus Ringnér, Carsten Peterson, Sofia Gruvberger, Yidong Chen, Sujatha Panavally, Lao H Saal, and Paul S Meltzer. 2001. "Estrogen Receptor Status in Breast Cancer Is Associated with Remarkably Distinct Gene Expression Patterns." *Cancer Research* 61 (16): 5979–84.

Gu, Xinjin, Chen Wang, Hui Deng, Chong Qing, Rong Liu, Sanhong Liu, and Xinying Xue. 2020. "Exosomal piRNA profiling revealed unique circulating piRNA signatures of cholangiocarcinoma and gallbladder carcinoma." *Acta biochimica et biophysica Sinica* 52 (5): 475-84. https://doi.org/10.1093/abbs/gmaa028.

Guinney, Justin, Rodrigo Dienstmann, Xin Wang, Aurélien De Reyniès, Andreas Schlicker, Charlotte Soneson, Laetitia Marisa, et al. 2015. "The Consensus Molecular Subtypes of Colorectal Cancer." *Nature Medicine* 21 (11): 1350–56. https://doi.org/10.1038/nm.3967.

Guo, Mingzhou, Yaojun Peng, Aiai Gao, Chen Du, and James G Herman. 2019. "Epigenetic Heterogeneity in Cancer." *Biomarker Research. BioMed Central* 7: 23. https://doi.org/10.1186/s40364-019-0174-y.

Hanna, Johora, Gazi S Hossain, and Jannet Kocerha. 2019. "The Potential for MicroRNA Therapeutics and Clinical Research." *Frontiers in Genetics. Frontiers* 10:478. https://doi.org/10.3389/fgene.2019.00478.

Hao, Zhao, Wei Fan, Jian Hao, Xin Wu, Guo Qing Zeng, Li Juan Zhang, Sui Feng Nie, and Xu Dong Wang. 2016. "Efficient Delivery of Micro RNA to Bone-Metastatic Prostate Tumors by Using Aptamer-Conjugated Atelocollagen in Vitro and in Vivo." *Drug Delivery* 23 (3): 874–81. https://doi.org/10.3109/10717544.2014.920059.

Harbeck, Nadia, Karl Sotlar, Rachel Wuerstlein, and Sophie Doisneau-Sixou. 2014. "Molecular and Protein Markers for Clinical Decision Making in Breast Cancer: Today and Tomorrow." *Cancer Treatment Reviews. Elsevier* 40(3), P434-44. https://doi.org/10.1016/j.ctrv.2013.09.014.

Harbers, Mathias, and Piero Carninci. 2005. "Tag-Based Approaches for Transcriptome Research and Genome Annotation." *Nature Methods. Nature Publishing Group* 2, 495–502. https://doi.org/10.1038/nmeth768.

Hayashizaki, Y. 2009. "Cap Analysis Gene Expression (CAGE)." *In Cap-Analysis Gene Expression (CAGE): The Science of Decoding Genes Transcription*, 49:1–5. https://doi.org/10.4032/9789814241359.

Heller, Michael J. 2002. "DNA Microarray Technology: Devices, Systems, and Applications." Annual Review of Biomedical Engineering. *Annual Reviews* 4(1): 129-53. https://doi.org/10.1146/annurev.bioeng.4.020702.153438.

Hemmatzadeh, Maryam, Hamed Mohammadi, Farhad Jadidi-Niaragh, Faezeh Asghari, and Mehdi Yousefi. 2016. "The Role of Oncomirs in the Pathogenesis and Treatment of Breast Cancer." *Biomedicine and Pharmacotherapy. Elsevier* 78:129-39 https://doi.org/10.1016/j.biopha.2016.01.026.

Hosseinahli, Nayer, Mahyar Aghapour, Pascal H.G. Duijf, and Behzad Baradaran. 2018. "Treating Cancer with MicroRNA Replacement Therapy: A Literature Review." *Journal of Cellular Physiology. Wiley Online Library* 233: 5574-88. https://doi.org/10.1002/jcp.26514.

Hu, Min, and Kornelia Polyak. 2006. "Serial Analysis of Gene Expression." *Nature Protocols* 1 (4): 1743–60. https://doi.org/10.1038/nprot.2006.269.

Huang, Shi-wei, Ni-da Ali, Lily Zhong, and Jian Shi.2018. "MicroRNAs as biomarkers for human glioblastoma: progress and potential." *Acta Pharmacologica Sinica* 39 (9): 1405-13. https://doi.org/10.1038/aps.2017.173.

Iliev, Robert, Michal Fedorko, Tana Machackova, Hana Mlcochova, Marek Svoboda, Dalibor Pacík, Jan Dolezel, Michal Stanik, and Ondrej Slaby. 2016. "Expression levels of PIWI-interacting RNA, piR-823, are deregulated in tumor tissue, blood serum and urine of patients with renal cell carcinoma." *Anticancer research* 36 (12): 6419-23.

Ishihara, Yoshitaka, Satoshi Tsuno, Satoshi Kuwamoto, Taro Yamashita, Yusuke Endo, Keigo Miura, Yugo Miura, Takemasa Sato, Junichi Hasegawa, and Norimasa Miura. "Correction to: Tumor-suppressive effects of atelocollagen-conjugated hsa-miR-520d-5p on un-differentiated cancer cells in a mouse xenograft model." *BMC cancer* 17 (1): 1-3.

Janiszewska, Michalina, Doris P Tabassum, Zafira Castaño, Simona Cristea, Kimiyo N Yamamoto, Natalie L Kingston, Katherine C Murphy, et al. 2019. "Subclonal Cooperation Drives Metastasis by Modulating Local and Systemic Immune Microenvironments." *Nature Cell Biology* 21 (7): 879–88. https://doi.org/10.1038/s41556-019-0346-x.

Kahraman, Mustafa, Anne Röske, Thomas Laufer, Tobias Fehlmann, Christina Backes, Fabian Kern, Jochen Kohlhaas et al. "MicroRNA in diagnosis and therapy monitoring of early-stage triple-negative breast cancer." *Scientific reports* 8 (1):1-11. https://doi.org/10.1038/s41598-018-29917-2.

Karlsson, Anna, Hans Brunnström, Patrick Micke, Srinivas Veerla, Johanna Mattsson, Linnea La Fleur, Johan Botling, et al. 2017. "Gene Expression Profiling of Large Cell Lung Cancer Links Transcriptional Phenotypes to the New Histological WHO 2015 Classification." *Journal of Thoracic Oncology* 12 (8): 1257–67. https://doi.org/10.1016/j.jtho.2017.05.008.

Katzenelenbogen, Yonatan, Fadi Sheban, Adam Yalin, Ido Yofe, Dmitry Svetlichnyy, Diego Adhemar Jaitin, Chamutal Bornstein, et al. 2020. "Coupled ScRNA-Seq and Intracellular Protein Activity Reveal an Immunosuppressive Role of TREM2 in Cancer." *Cell* 182 (4): 872-885.e19. https://doi.org/10.1016/j.cell.2020.06.032.

Khan, Javed, Jun S Wei, Markus Ringnér, Lao H Saal, Marc Ladanyi, Frank Westermann, Frank Berthold, et al. 2001. "Classification and Diagnostic Prediction of Cancers Using Gene Expression Profiling and Artificial Neural Networks." *Nature Medicine* 7 (6): 673–79. https://doi.org/10.1038/89044.

Kim, Sang Wook. 2017. "The Role of MicroRNAs in Colorectal Cancer." *The Korean Journal of Gastroenterology = Taehan Sohwagi Hakhoe Chi. NIH Public Access* 69 (4): 206-11. https://doi.org/10.4166/kjg.2017.69.4.206.

Kloosterman, Wigard P, Robert R. J. Coebergh Van Den Braak, Mark Pieterse, Markus J Van Roosmalen, Anieta M Sieuwerts, Christina Stangl, Ronne Brunekreef, et al. 2017. "A Systematic Analysis of Oncogenic Gene Fusions in Primary Colon Cancer." *Cancer Research* 77 (14): 3814–22. https://doi.org/10.1158/0008-5472.CAN-16-3563.

Knezevic, Dejan, Audrey D Goddard, Nisha Natraj, Diana B Cherbavaz, Kim M Clark-Langone, Jay Snable, Drew Watson, et al. 2013. "Analytical Validation of the Oncotype DX Prostate Cancer Assay - a Clinical RT-PCR Assay Optimized for Prostate Needle Biopsies." *BMC Genomics* 14 (1): 1–12. https://doi.org/10.1186/1471-2164-14-690.

Kodzius, Rimantas, Miki Kojima, Hiromi Nishiyori, Mari Nakamura, Shiro Fukuda, Michihira Tagami, Daisuke Sasaki, et al. 2006. "Cage: Cap Analysis of Gene Expression." *Nature Methods* 3 (3): 211. https://doi.org/10.1038/nmeth0306-211.

Kondov, Borislav, Zvonko Milenkovikj, Goran Kondov, Gordana Petrushevska, Neli Basheska, Magdalena Bogdanovska-Todorovska, Natasha Tolevska, and Ljube Ivkovski. 2018. "Presentation of the Molecular Subtypes of Breast Cancer Detected by Immunohistochemistry in Surgically Treated Patients." *Open Access Macedonian Journal of Medical Sciences* 6 (6): 961–67. https://doi.org/10.3889/oamjms.2018.231.

Kopetz, Scott, and James L Abbruzzese. 2009. "Barriers to Integrating Gene Profiling for Stage II Colon Cancer." *Clinical Cancer Research* 15 (24): 7451–52. https://doi.org/ 10.1158/1078-0432.CCR-09-2523.

Kosanam, Hari, Ioannis Prassas, Caitlin C. Chrystoja, Ireena Soleas, Alison Chan, Apostolos Dimitromanolakis, Ivan M. Blasutig et al. 2013. "Laminin, gamma 2 (LAMC2): a promising new putative pancreatic cancer biomarker identified by proteomic analysis of pancreatic adenocarcinoma tissues." *Molecular & Cellular Proteomics* 12 (10): 2820-32. doi:10.1074/mcp.M112.023507.

Labidi-Galy, S Intidhar, Eniko Papp, Dorothy Hallberg, Noushin Niknafs, Vilmos Adleff, Michael Noe, Rohit Bhattacharya, et al. 2017. "High Grade Serous Ovarian Carcinomas Originate in the Fallopian Tube." *Nature Communications* 8 (1): 1–11. https://doi.org/10.1038/s41467-017-00962-1.

Lennox, Kim A, Richard Owczarzy, Derek M Thomas, Joseph A Walder, and Mark A Behlke. 2013. "Improved Performance of Anti-MiRNA Oligonucleotides Using a Novel Non-Nucleotide Modifier." *Molecular Therapy - Nucleic Acids* 2: e117. https://doi.org/10.1038/mtna.2013.46.

Lenz, Heinz Josef, Fang Shu Ou, Alan P Venook, Howard S Hochster, Donna Niedzwiecki, Richard M Goldberg, Robert J Mayer, et al. 2019. "Impact of Consensus Molecular Subtype on Survival in Patients with Metastatic Colorectal Cancer: Results from CALGB/SWOG 80405 (Alliance)." *In Journal of Clinical Oncology*, 37:1876–85. American Society of Clinical Oncology. https://doi.org/10. 1200/JCO.18.02258.

Li, Beibei, Jiaxin Hong, Mei Hong, Yajun Wang, Tingting Yu, Sibin Zang, and Qiuling Wu. 2019. "piRNA-823 delivered by multiple myeloma-derived extracellular vesicles promoted tumorigenesis through re-educating endothelial cells in the tumor environment." *Oncogene* 38(26): 5227-38. https://doi.org/10.1038/s41388-019-0788-4.

Li, Xiang, and Kong Beihua. 2013. "PAX8 Is a Novel Marker for Differentiating between Various Types of Tumor, Particularly Ovarian Epithelial Carcinomas (Review)." *Oncology Letters. Spandidos Publications* 5(3): 735-38. https://doi.org/10.3892/ol. 2013.1121.

Li, Yuping, Xiwei Wu, Hanlin Gao, Jennifer M Jin, Arthur X Li, Young S Kim, Sumanta K Pal, et al. 2015. "Piwi-Interacting RNAs (PiRNAs) Are Dysregulated in Renal Cell Carcinoma and Associated with Tumor Metastasis and Cancer-Specific Survival." *Molecular Medicine* 21 (1): 381–88. https://doi.org/10.2119/molmed. 2014.00203.

Liao, Jipei, Lei Yu, Yuping Mei, Maria Guarnera, Jun Shen, Ruiyun Li, Zhenqiu Liu, and Feng Jiang. 2010. "Small Nucleolar RNA Signatures as Biomarkers for Non-Small-Cell Lung Cancer." *Molecular Cancer* 9 (1): 1–10. https://doi.org/10.1186/1476-4598-9-198.

Lima, Joana Filipa, Laura Cerqueira, Ceu Figueiredo, Carla Oliveira, and Nuno Filipe Azevedo. 2018. "Anti-MiRNA Oligonucleotides: A Comprehensive Guide for Design." *RNA Biology. Taylor & Francis* 15 (3): 338-52. https://doi.org/10.1080/15476286.2018.1445959.

Lister, Ryan, Ronan C O'Malley, Julian Tonti-Filippini, Brian D Gregory, Charles C Berry, A Harvey Millar, and Joseph R Ecker. 2008. "Highly Integrated Single-Base Resolution Maps of the Epigenome in Arabidopsis." *Cell* 133 (3): 523–36. https://doi.org/10.1016/j.cell.2008.03.029.

Liu, Hairong, Juan Li, Pratirodh Koirala, Xianfeng Ding, Binghai Chen, Yiheng Wang, Zheng Wang, Chuanxin Wang, Xu Zhang, and Yin Yuan Mo. 2016. "Long Non-Coding RNAs as Prognostic Markers in Human Breast Cancer." *Oncotarget* 7 (15): 20584–96. https://doi.org/10.18632/oncotarget.7828.

Liu, Rui, Chunni Zhang, Zhibin Hu, Gou Li, Cheng Wang, Cuihua Yang, Dingzhi Huang, et al. 2011. "A Five-MicroRNA Signature Identified from Genome-Wide Serum MicroRNA Expression Profiling Serves as a Fingerprint for Gastric Cancer Diagnosis." *European Journal of Cancer* 47 (5): 784–91. https://doi.org/10.1016/j.ejca.2010.10.025.

Liu, Yongmei, Mei Dou, Xuxia Song, Yanhan Dong, Si Liu, Haoran Liu, Jiaping Tao, Wenjing Li, Xunhua Yin, and Wenhua Xu. 2019. "The Emerging Role of the PiRNA/Piwi Complex in Cancer." *Molecular Cancer. BioMed Central* 18: 123. https://doi.org/10.1186/s12943-019-1052-9.

Loi, Sherene, Benjamin Haibe-Kains, Christine Desmedt, Françoise Lallemand, Andrew M Tutt, Cheryl Gillet, Paul Ellis, et al. 2007. "Definition of Clinically Distinct Molecular Subtypes in Estrogen Receptor-Positive Breast Carcinomas through Genomic Grade." *Journal of Clinical Oncology* 25 (10): 1239–46. https://doi.org/10.1200/JCO.2006.07.1522.

Lowe, Rohan, Neil Shirley, Mark Bleackley, Stephen Dolan, and Thomas Shafee. 2017. "Transcriptomics Technologies." *PLoS Computational Biology* 13 (5): e1005457. https://doi.org/10.1371/journal.pcbi.1005457.

Lu, Qin, Ting Yu, Xilong Ou, Dazhong Cao, Ting Xie, and Xia Chen. 2017. "Potential LncRNA Diagnostic Biomarkers for Early Gastric Cancer." *Molecular Medicine Reports* 16 (6): 9545–52. https://doi.org/10.3892/mmr.2017.7770.

Maak, Matthias, Iris Simon, Ulrich Nitsche, Paul Roepman, Mireille Snel, Annuska M Glas, Tibor Schuster, et al. 2013. "Independent Validation of a Prognostic Genomic Signature (ColoPrint) for Patients with Stage II Colon Cancer." *Annals of Surgery* 257 (6): 1053–58. https://doi.org/10.1097/SLA.0b013e31827c1180.

Macosko, Evan Z, Anindita Basu, Rahul Satija, James Nemesh, Karthik Shekhar, Melissa Goldman, Itay Tirosh, et al. 2015. "Highly Parallel Genome-Wide Expression Profiling of Individual Cells Using Nanoliter Droplets." *Cell* 161 (5): 1202–14. https://doi.org/10.1016/j.cell.2015.05.002.

Mahapatra, Kunal Das, Lorenzo Pasquali, Jonas Nørskov Søndergaard, Jan Lapins, István Balazs Nemeth, Eszter Baltás, Lajos Kemény, et al. 2020. "A Comprehensive Analysis of Coding and Non-Coding Transcriptomic Changes in Cutaneous Squamous Cell Carcinoma." *Scientific Reports* 10 (1): 1–12. https://doi.org/10. 1038/s41598-020-59660-6.

Marisa, Laetitia, Mira Ayadi, Ralyath Balogoun, Camilla Pilati, Karine Le Malicot, Come Lepage, Jean-François Emile, et al. 2017. "Clinical Utility of Colon Cancer Molecular Subtypes: Validation of Two Main Colorectal Molecular Classifications on the PETACC-8 Phase III Trial Cohort." *Journal of Clinical Oncology* 35 (15_suppl): 3509. https://doi.org/10.1200/jco.2017.35.15_suppl.3509.

Marra, Marco A, Ladeana Hillier, and Robert H Waterston. 1998. "Expressed Sequence Tags--ESTablishing Bridges between Genomes." *Trends in Genetics* : TIG 14 (1): 4-7. https://doi.org/10.1016/S0168-9525(97)01355-3.

Maynard, Ashley, Caroline E McCoach, Julia K Rotow, Lincoln Harris, Franziska Haderk, D Lucas Kerr, Elizabeth A. Yu, et al. 2020. "Therapy-Induced Evolution of Human Lung Cancer Revealed by Single-Cell RNA Sequencing." *Cell* 182 (5): 1232-51.e22. https://doi.org/10.1016/j.cell.2020.07.017.

McGettigan, Paul A. 2013. "Transcriptomics in the RNA-Seq Era." *Current Opinion in Chemical Biology. Elsevier* 17(1):4-11. https://doi.org/10.1016/j.cbpa.2012.12.008.

Mercer, Tim R, Michael B Clark, Joanna Crawford, Marion E Brunck, Daniel J Gerhardt, Ryan J Taft, Lars K Nielsen, Marcel E Dinger, and John S Mattick. 2014. "Targeted Sequencing for Gene Discovery and Quantification Using RNA CaptureSeq." *Nature Protocols* 9 (5): 989-1009. https://doi.org/10.1038/nprot.2014.058.

Mollaei, Homa, Reza Safaralizadeh, and Zeinab Rostami. 2019. "MicroRNA Replacement Therapy in Cancer." *Journal of Cellular Physiology. Wiley Online Library* 234(8):12369-84. https://doi.org/10.1002/jcp.28058.

Mooi, Jennifer K, P Wirapati, R Asher, C K Lee, P Savas, T J Price, A Townsend, et al. 2018. "The Prognostic Impact of Consensus Molecular Subtypes (CMS) and Its Predictive Effects for Bevacizumab Benefit in Metastatic Colorectal Cancer: Molecular Analysis of the AGITG MAX Clinical Trial." *Annals of Oncology* 29 (11): 2240-46. https://doi.org/10.1093/annonc/mdy410.

Mosig, Rebecca A, Mollie Lobl, Emir Senturk, Hardik Shah, Samantha Cohen, Eugene Chudin, Robert Fruscio, et al. 2012. "IGFBP-4 Tumor and Serum Levels Are Increased across All Stages of Epithelial Ovarian Cancer." *Journal of Ovarian Research* 5 (1): 1-8. https://doi.org/10.1186/1757-2215-5-3.

Moss, E L, J Hollingworth, and T M Reynolds. 2005. "The Role of CA125 in Clinical Practice." *Journal of Clinical Pathology* 58 (3): 308-12. https://doi.org/10.1136/jcp.2004.018077.

Mulcahy Nick. "It's Oncotype DX vs MammaPrint in Europe." *Medscape, 2014*. (cited 2014 May 8). Available from: http://www.medscape.com/viewarticle/824874.

Nadal, Ernest, Anna Truini, Asuka Nakata, Jules Lin, Rishindra M Reddy, Andrew C Chang, Nithya Ramnath, Noriko Gotoh, David G Beer, and Guoan Chen. 2015. "A Novel Serum 4-MicroRNA Signature for Lung Cancer Detection." *Scientific Reports* 5 (1): 1-9. https://doi.org/10.1038/srep12464.

Nassri, Ammar, Hong Zhu, Mayssan Muftah, and Zeeshan Ramzan. 2018. "Epidemiology and Survival of Esophageal Cancer Patients in an American Cohort." *Cureus* 10 (4): e2507. https://doi.org/10.7759/cureus.2507.

Nayak, Sushrusha, and Roland W Herzog. 2010. "Progress and Prospects: Immune Responses to Viral Vectors." *Gene Therapy. Nature Publishing Group* 17, 295–304. https://doi.org/10.1038/gt.2009.148.

Nedaeinia, Reza, Amir Avan, Mehdi Ahmadian, Sasan Nedaee Nia, Maryam Ranjbar, Mohammadreza Sharifi, Mohammad Goli, et al. 2017. "Current Status and Perspectives Regarding LNA-Anti-MiR Oligonucleotides and MicroRNA MiR-21 Inhibitors as a Potential Therapeutic Option in Treatment of Colorectal Cancer." *Journal of Cellular Biochemistry* 118 (12): 4129-40. https://doi.org/10.1002/jcb.26047.

Nelson, Nancy J. 2001. "Microarrays Have Arrived: Gene Expression Tool Matures." *Journal of the National Cancer Institute. Oxford University Press* 93 (7) 492-94. https://doi.org/10.1093/jnci/93.7.492.

O'Neill, Clodagh P., and Róisín M Dwyer. 2020. "Nanoparticle-Based Delivery of Tumor Suppressor MicroRNA for Cancer Therapy." *Cells. Multidisciplinary Digital Publishing Institute* 9(2): 521. https://doi.org/10.3390/cells9020521.

Okita, Akira, Shin Takahashi, Kota Ouchi, Masahiro Inoue, Mika Watanabe, Mareyuki Endo, Hiroshi Honda, Yasuhide Yamada, and Chikashi Ishioka. 2018. "Consensus Molecular Subtypes Classification of Colorectal Cancer as a Predictive Factor for Chemotherapeutic Efficacy against Metastatic Colorectal Cancer." *Oncotarget* 9 (27): 18698–711. https://doi.org/10.18632/oncotarget.24617.

Okoniewski, Michał J, and Crispin J Miller. 2006. "Hybridization Interactions between Probesets in Short Oligo Microarrays Lead to Spurious Correlations." *BMC Bioinformatics* 7 (1): 1-14. https://doi.org/10.1186/1471-2105-7-276.

Osaki, Mitsuhiko, Fumitaka Takeshita, Yui Sugimoto, Nobuyoshi Kosaka, Yusuke Yamamoto, Yusuke Yoshioka, Eisuke Kobayashi, et al. 2011. "MicroRNA-143 Regulates Human Osteosarcoma Metastasis by Regulating Matrix Metalloprotease-13 Expression." *Molecular Therapy* 19 (6): 1123-30. https://doi.org/10.1038/mt.2011.53.

Paik, Soonmyung, Gong Tang, Steven Shak, Chungyeul Kim, Joffre Baker, Wanseop Kim, Maureen Cronin, et al. 2006. "Gene Expression and Benefit of Chemotherapy in Women with Node-Negative, Estrogen Receptor-Positive Breast Cancer." *Journal of Clinical Oncology* 24 (23): 3726-34. https://doi.org/10.1200/JCO.2005.04.7985.

Pan, Qun, Ofer Shai, Leo J Lee, Brendan J Frey, and Benjamin J Blencowe. 2009. "Erratum: Deep Surveying of Alternative Splicing Complexity in the Human Transcriptome by High-Throughput Sequencing (Nature Genetics (2008) 40 (1413-1415)." *Nature Genetics. Nature Publishing Group* 41, 762. https://doi.org/10.1038/ng0609-762d.

Pan, Yu, Fengchun Lu, Qinglin Fei, Xingxing Yu, Ping Xiong, Xunbin Yu, Yuan Dang, et al. 2019. "Single-Cell RNA Sequencing Reveals Compartmental Remodeling of Tumor-Infiltrating Immune Cells Induced by Anti-CD47 Targeting in Pancreatic Cancer." *Journal of Hematology and Oncology* 12 (1): 1-18. https://doi.org/10.1186/s13045-019-0822-6.

Peiffer, Jason A, Shail Kaushik, Hajime Sakai, Mario Arteaga-Vazquez, Nidia Sanchez-Leon, Hassan Ghazal, Jean Philippe Vielle-Calzada, and Blake C Meyers. 2008. "A Spatial Dissection of the Arabidopsis Floral Transcriptome by MPSS." *BMC Plant Biology* 8 (1): 1-16. https://doi.org/10.1186/1471-2229-8-43.

Pepin, Kristen, Marcela del Carmen, Amy Brown, and Don S Dizon. 2014. "CA 125 and Epithelial Ovarian Cancer: Role in Screening, Diagnosis, and Surveillance." *American Journal of Hematology / Oncology* 10 (6): 22-29. http://gotoper-com.s3.amazonaws.com/_media/_pdf/AJHO_Dec14_03_Ovarian.pdf.

Peres, Lauren C, Kara L Cushing-Haugen, Michael Anglesio, Kristine Wicklund, Rex Bentley, Andrew Berchuck, Linda E Kelemen, et al. 2018. "Histotype Classification of Ovarian Carcinoma: A Comparison of Approaches." *Gynecologic Oncology* 151 (1): 53–60. https://doi.org/10.1016/j.ygyno.2018.08.016.

Perou, Charles M, Therese Sørile, Michael B Eisen, Matt Van De Rijn, Stefanie S Jeffrey, Christian A. Ress, Jonathan R Pollack, et al. 2000. "Molecular Portraits of Human Breast Tumours." *Nature* 406 (6797): 747-52. https://doi.org/10.1038/35021093.

Pozhitkov, Alex E, Diethard Tautz, and Peter A Noble. 2007. "Oligonucleotide Microarrays: Widely Applied-Poorly Understood." *Briefings in Functional Genomics and Proteomics* 6 (2): 141-48. https://doi.org/10.1093/bfgp/elm014.

Putney, Scott D, Walter C Herlihy, and Paul Schimmel. 1983. "A New Troponin T and CDNA Clones for 13 Different Muscle Proteins, Found by Shotgun Sequencing." *Nature. Nature Publishing Group* 302, 718–21. https://doi.org/10.1038/302718a0.

Qiu, Yunping, Guoxiang Cai, Bingsen Zhou, Dan Li, Aihua Zhao, Guoxiang Xie, Houkai Li, et al. 2014. "A Distinct Metabolic Signature of Human Colorectal Cancer with Prognostic Potential." *Clinical Cancer Research* 20 (8): 2136–46. https://doi.org/10.1158/1078-0432.CCR-13-1939.

Reimers, Marlies S., Peter J. K. Kuppen, Mark Lee, Margarita Lopatin, Haluk Tezcan, Hein Putter, Kim Clark-Langone, Gerrit J.an Liefers, Steven Shak, and Cornelis J. H. van de Velde. 2014. "Validation of the 12-Gene Colon Cancer Recurrence Score as a Predictor of Recurrence Risk in Stage II and III Rectal Cancer Patients." *Journal of the National Cancer Institute* 106 (11): 4512. https://doi.org/10.1093/jnci/dju269.

Reinartz, Jeannette, Eddy Bruyns, Jing Zhong Lin, Tim Burcham, Sydney Brenner, Ben Bowen, Michael Kramer, and Rick Woychik. 2002. "Technique Review Massively Parallel Signature Sequencing (MPSS) as a Tool for in-Depth Quantitative Gene Expression Profiling in All Organisms." *Briefings in Functional Genomics and Proteomics. Henry Stewart Publications* 1(1): 95-104. https://doi.org/10.1093/bfgp/1.1.95.

Reis-Filho, Jorge S, and Lajos Pusztai. 2011. "Gene Expression Profiling in Breast Cancer: Classification, Prognostication, and Prediction." *The Lancet.* Elsevier. 378(9805): P1812-23. https://doi.org/10.1016/S0140-6736(11)61539-0.

Ren, Shancheng, Zhiyu Peng, Jian Hua Mao, Yongwei Yu, Changjun Yin, Xin Gao, Zilian Cui, et al. 2012. "RNA-Seq Analysis of Prostate Cancer in the Chinese Population Identifies Recurrent Gene Fusions, Cancer-Associated Long Noncoding RNAs and Aberrant Alternative Splicings." *Cell Research* 22 (5): 806-21. https://doi.org/10.1038/cr.2012.30.

Rizzo, Francesca, Antonio Rinaldi, Giovanna Marchese, Elena Coviello, Assunta Sellitto, Angela Cordella, Giorgio Giurato, et al. 2016. "Specific Patterns of PIWI-Interacting Small Noncoding RNA Expression in Dysplastic Liver Nodules and Hepatocellular Carcinoma." *Oncotarget* 7 (34): 54650-61. https://doi.org/10.18632/oncotarget.10567.

Robinson, Mark D, and Alicia Oshlack. 2010. "A Scaling Normalization Method for Differential Expression Analysis of RNA-Seq Data." *Genome Biology* 11 (3): 1-9. https://doi.org/10.1186/gb-2010-11-3-r25.

Rodgers, Laura H, Eoghainín HAinmhire, Alexandria N. Young, and Joanna E Burdette. 2016. "Loss of PAX8 in High-Grade Serous Ovarian Cancer Reduces Cell Survival despite Unique Modes of Action in the Fallopian Tube and Ovarian Surface Epithelium." *Oncotarget* 7 (22): 32785-95. https://doi.org/10.18632/oncotarget.9051.

Rouzier, Roman, Charles M Perou, W Fraser Symmans, Nuhad Ibrahim, Massimo Cristofanilli, Keith Anderson, Kenneth R Hess, et al. 2005. "Breast Cancer Molecular Subtypes Respond Differently to Preoperative Chemotherapy." *Clinical Cancer Research* 11 (16): 5678-85. https://doi.org/10.1158/1078-0432.CCR-04-2421.

Royce, Thomas E, Joel S Rozowsky, and Mark B Gerstein. 2007. "Toward a Universal Microarray: Prediction of Gene Expression through Nearest-Neighbor Probe Sequence Identification." *Nucleic Acids Research* 35 (15): e99. https://doi.org/10.1093/nar/gkm549.

Saldanha, Gerald, Linda Potter, Priya Shendge, Joy Osborne, Steve Nicholson, NgiWieh Yii, Sanjay Varma et al. 2013. "Plasma microRNA-21 is associated with tumor burden in cutaneous melanoma." *J Invest Dermatol* 133(5):1381-84. https://doi.org/10.1038/jid.2012.477.

Salem, Hanaa, Gamal Attiya, and Nawal El-Fishawy. 2017. "Classification of Human Cancer Diseases by Gene Expression Profiles." *Applied Soft Computing Journal* 50: 124-34. https://doi.org/10.1016/j.asoc.2016.11.026.

Sandhu, Rupninder, Joel S Parker, Wendell D Jones, Chad A Livasy, and William B Coleman. 2010. "Microarray-Based Gene Expression Profiling for Molecular Classification of Breast Cancer and Identification of New Targets for Therapy." *Laboratory Medicine.* Oxford University Press Oxford, UK 41(6):364–72. https://doi.org/10.1309/LMLIK0VIE3CJK0WD.

Schade, Anna, Evgenya Delyagina, Dorothee Scharfenberg, Anna Skorska, Cornelia Lux, Robert David, and Gustav Steinhoff. 2013. "Innovative Strategy for MicroRNA Delivery in Human Mesenchymal Stem Cells via Magnetic Nanoparticles." *International Journal of Molecular Sciences* 14 (6): 10710-26. https://doi.org/10.3390/ijms140610710.

Schena, Mark, Dari Shalon, Ronald W Davis, and Patrick O Brown. 1995. "Quantitative Monitoring of Gene Expression Patterns with a Complementary DNA Microarray." *Science* 270 (5235): 467-70. https://doi.org/10.1126/science.270.5235.467.

Selewa, Alan, Ryan Dohn, Heather Eckart, Stephanie Lozano, Bingqing Xie, Eric Gauchat, Reem Elorbany, Katherine Rhodes, Jonathan Burnett, and Yoav Gilad. 2020. "Systematic Comparison of High-Throughput Single-Cell and Single-Nucleus Transcriptomes during Cardiomyocyte Differentiation." *Scientific Reports* 10 (1): 1-13. https://doi.org/10.1038/s41598-020-58327-6.

Seto, Anita G., Xuan Beatty, Joshua M. Lynch, Melanie Hermreck, Michael Tetzlaff, Madeleine Duvic, and Aimee L. Jackson. 2018. "Cobomarsen, an oligonucleotide inhibitor of miR-155, co-ordinately regulates multiple survival pathways to reduce cellular proliferation and survival in cutaneous T-cell lymphoma." *British journal of haematology* 183,: 428-44. https://doi.org/10.1111/bjh.15547.

Shalek, Alex K, Rahul Satija, Xian Adiconis, Rona S Gertner, Jellert T Gaublomme, Raktima Raychowdhury, Schraga Schwartz, et al. 2013. "Single-Cell Transcriptomics Reveals Bimodality in Expression and Splicing in Immune Cells." *Nature* 498 (7453): 236-40. https://doi.org/10.1038/nature12172.

Sharma, Ankur, Elaine Yiqun Cao, Vibhor Kumar, Xiaoqian Zhang, Hui Sun Leong, Angeline Mei Lin Wong, Neeraja Ramakrishnan, et al. 2018. "Longitudinal Single-Cell RNA Sequencing of Patient-Derived Primary Cells Reveals Drug-Induced Infidelity in Stem Cell Hierarchy." *Nature Communications* 9 (1): 1-17. https://doi.org/10.1038/s41467-018-07261-3.

Shen, Yao, Arianna L Kim, Rong Du, and Liang Liu. 2016. "Transcriptome Analysis Identifies the Dysregulation of Ultraviolet Target Genes in Human Skin Cancers." *PLoS ONE* 11 (9): e0163054. https://doi.org/10.1371/journal.pone.0163054.

Shiraki, Toshiyuki, Shinji Kondo, Shintaro Katayama, Kazunori Waki, Takeya Kasukawa, Hideya Kawaji, Rimantas Kodzius, et al. 2003. "Cap Analysis Gene Expression for High-Throughput Analysis of Transcriptional Starting Point and Identification of Promoter Usage." *Proceedings of the National Academy of Sciences of the United States of America* 100 (26): 15776-81. https://doi.org/10.1073/pnas. 2136655100.

Sim, G K, F C Kafatos, C W Jones, M D Koehler, A Efstratiadis, and T Maniatis. 1979. "Use of a CDNA Library for Studies on Evolution and Developmental Expression of the Chorion Multigene Families." *Cell* 18 (4): 1303-16. https://doi.org/10.1016/ 0092-8674(79)90241-1.

Slaby, Ondrej. 2016. "Non-Coding RNAs as Biomarkers for Colorectal Cancer Screening and Early Detection." *Advances in Experimental Medicine and Biology* 937: 153-70. https://doi.org/10.1007/978-3-319-42059-2_8.

Solé, Carla, Daniela Tramonti, Maike Schramm, Ibai Goicoechea, María Armesto, Luiza I. Hernandez, Lorea Manterola, Marta Fernandez-Mercado, Karmele Mujika, Anna Tuneu, Ane Jaka, Maitena Tellaetxe, Marc R. Friedländer, Xavier Estivill, Paolo Piazza, Pablo L. Ortiz-Romero, Mark R. Middleton, and Charles H. Lawrie. 2019. "The Circulating Transcriptome as a Source of Biomarkers for Melanoma" Cancers 11(1): 70. https://doi.org/10.3390/cancers11010070.

Solin, Lawrence J, Robert Gray, Frederick L Baehner, Steven M Butler, Lorie L Hughes, Carl Yoshizawa, Diana B Cherbavaz, et al. 2013. "A Multigene Expression Assay to Predict Local Recurrence Risk for Ductal Carcinoma in Situ of the Breast." *Journal of the National Cancer Institute* 105 (10): 701-10. https://doi.org/10.1093/jnci/djt067.

Sørlie, Therese, Charles M Perou, Robert Tibshirani, Turid Aas, Stephanie Geisler, Hilde Johnsen, Trevor Hastie, et al. 2001. "Gene Expression Patterns of Breast Carcinomas Distinguish Tumor Subclasses with Clinical Implications." *Proceedings of the National Academy of Sciences of the United States of America* 98 (19): 10869-74. https://doi.org/10.1073/pnas.191367098.

Stebbing, Justin, and Harpreet Singh Wasan. 2019. "Decoding Metastatic Colorectal Cancer to Improve Clinical Decision Making." *Journal of Clinical Oncology. American Society of Clinical Oncology* 37(22):1847-50. https://doi.org/10.1200/JCO. 19.01185.

Stenvang, Jan, Andreas Petri, Morten Lindow, Susanna Obad, and Sakari Kauppinen. 2012. "Inhibition of MicroRNA Function by AntimiR Oligonucleotides." *Silence. BioMed Central* 3(1): 1-17. https://doi.org/10.1186/1758-907X-3-1.

Stintzing, Sebastian, Pratyaksha Wirapati, H. J. Lenz, Daniel Neureiter, Ludwig Fischer von Weikersthal, Thomas Decker, Alexander Kiani, et al. 2019. "Consensus Molecular Subgroups (CMS) of Colorectal Cancer (CRC) and First-Line Efficacy of FOLFIRI plus Cetuximab or Bevacizumab in the FIRE3 (AIO KRK-0306) Trial." *Annals of Oncology* 30 (11): 1796-1803. https://doi.org/10.1093/annonc/mdz387.

Sun, Yu Meng, and Yue Qin Chen. 2020. "Principles and Innovative Technologies for Decrypting Noncoding RNAs: From Discovery and Functional Prediction to Clinical Application." *Journal of Hematology and Oncology. Springer* 13 (1): 1-27. https://doi.org/10.1186/s13045-020-00945-8.

Sutcliffe, J Gregor, Robert J Milner, Floyd E Bloom, and Richard A Lerner. 1982. "Common 82-Nucleotide Sequence Unique to Brain RNA." *Proceedings of the National Academy of Sciences of the United States of America* 79 (16 I): 4942-46. https://doi.org/10.1073/pnas.79.16.4942.

Thermes, Claude. 2014. "Ten Years of Next-Generation Sequencing Technology." *Trends in Genetics : TIG.* Elsevier. 30 (9): P418-426. https://doi.org/10.1016/j.tig.2014.07.001.

Tivnan, Amanda, Wayne Shannon Orr, Vladimir Gubala, Robert Nooney, David E Williams, Colette McDonagh, Suzanne Prenter, et al. 2012. "Inhibition of Neuroblastoma Tumor Growth by Targeted Delivery of MicroRNA-34a Using Anti-Disialoganglioside GD2 Coated Nanoparticles." *PLoS ONE* 7 (5): e38129. https://doi.org/10.1371/journal.pone.0038129.

Toffalorio, F, E Belloni, M Barberis, G Bucci, L Tizzoni, G Pruneri, C Fumagalli, et al. 2014. "Gene Expression Profiling Reveals GC and CEACAM1 as New Tools in the Diagnosis of Lung Carcinoids." *British Journal of Cancer* 110 (5): 1244-49. https://doi.org/10.1038/bjc.2014.41.

Toiyama, Yuji, Yoshinaga Okugawa, and Ajay Goel. 2014. "DNA Methylation and MicroRNA Biomarkers for Noninvasive Detection of Gastric and Colorectal Cancer." *Biochemical and Biophysical Research Communications. Elsevier* 455 (1-2): 43-57. https://doi.org/10.1016/j.bbrc.2014.08.001.

Tschentscher, Frank, Johannes Hüsing, Tanja Hölter, Elisabeth Kruse, Irina Gana Dresen, Karl Heinz Jöckel, Gerasimos Anastassiou, et al. 2003. "Tumor Classification Based on Gene Expression Profiling Shows That Uveal Melanomas with and without Monosomy 3 Represent Two Distinct Entities." *Cancer Research* 63 (10): 2578-84.

Tsujiura, M., Ichikawa, D., Komatsu, S. et al. "Circulating microRNAs in plasma of patients with gastric cancers." 2010. *British Journal of Cancer* 102, 1174-79. https://doi.org/10.1038/sj.bjc.6605608.

Vickers, Kasey C, and Alan T Remaley. 2012. "Lipid-Based Carriers of MicroRNAs and Intercellular Communication." *Current Opinion in Lipidology. NIH Public Access* 23 (2) 91-97. https://doi.org/10.1097/MOL.0b013e328350a425.

Vychytilova-Faltejskova, Petra, Karolina Stitkovcova, Lenka Radova, Milana Sachlova, Zdenka Kosarova, Katerina Slaba, Zdenek Kala, et al. 2018. "Circulating PIWI-Interacting RNAs PiR-5937 and PiR-28876 Are Promising Diagnostic Biomarkers of Colon Cancer." *Cancer Epidemiology Biomarkers and Prevention* 27 (9): 1019-28. https://doi.org/10.1158/1055-9965.EPI-18-0318.

Wang, Hairong, Shiming Liu, Li Jia, Fengyun Chu, Ya Zhou, Zhixu He, Mengmeng Guo, Chao Chen, and Lin Xu. 2018. "Nanostructured Lipid Carriers for MicroRNA Delivery in Tumor Gene Therapy." *Cancer Cell International. BioMed Central* 18 (1): 1-6. https://doi.org/10.1186/s12935-018-0596-x.

Wang, Haiying, Julian Molina, John Jiang, Matthew Ferber, Sandhya Pruthi, Timothy Jatkoe, Carlo Derecho, Yashoda Rajpurohit, Jian Zheng, and Yixin Wang 2013. "Gene expression markers in circulating tumor cells may predict bone metastasis and response to hormonal treatment in breast cancer." *Molecular and Clinical Oncology* 1 (6):1031-1038. https://doi.org/10.3892/mco.2013.163.

Wang, Jinglu, Dylan C Dean, Francis J Hornicek, Huirong Shi, and Zhenfeng Duan. 2019. "RNA Sequencing (RNA-Seq) and Its Application in Ovarian Cancer." *Gynecologic Oncology.* Elsevier 152 (1): P194-201. https://doi.org/10.1016/j.ygyno. 2018.10.002.

Wang, Lihong, Jia Ge, Yang Lan, Yu Shi, Ying Luo, Yuhuan Tan, Mei Liang, et al. 2020. "Tumor Mutational Burden Is Associated with Poor Outcomes in Diffuse Glioma." *BMC Cancer* 20 (1): 1-12. https://doi.org/10.1186/s12885-020-6658-1.

Wang, Ruifen, Yeqi Sun, Wenwei Yu, Yu Yan, Meng Qiao, Ruiqi Jiang, Wenbin Guan, and Lifeng Wang. 2019. "Downregulation of MiRNA-214 in Cancer-Associated Fibroblasts Contributes to Migration and Invasion of Gastric Cancer Cells through Targeting FGF9 and Inducing EMT." *Journal of Experimental and Clinical Cancer Research* 38 (1): 1-15. https://doi.org/10.1186/s13046-018-0995-9.

Wang, Wen Tao, Cai Han, Yu Meng Sun, Tian Qi Chen, and Yue Qin Chen. 2019. "Noncoding RNAs in Cancer Therapy Resistance and Targeted Drug Development." *Journal of Hematology and Oncology. BioMed Central* 12 (1): 1-15. https://doi.org/ 10.1186/s13045-019-0748-z.

Wang, Zhong, Mark Gerstein, and Michael Snyder. 2009. "RNA-Seq: A Revolutionary Tool for Transcriptomics." *Nature Reviews Genetics. Nature Publishing Group* 10 (1): 57-63. https://doi.org/10.1038/nrg2484.

Warf, M. Bryan, Placede G. Fosso, Elisha Hughes, Michael Perry, Krystal L. Brown, Julia E. Reid, Kathryn A. Kolquist, Susanne Wagner, Alexander Gutin, and Benjamin Roa. 2015. "Analytical Validation of a Proliferation-Based Molecular Signature Used as a Prognostic Marker in Early Stage Lung Adenocarcinoma." *Biomarkers in Medicine* 9 (9): 901-10. https://doi.org/10.2217/bmm.15.46.

Weiler, Jan, Juerg Hunziker, and Jeffrey Hall. 2006. "Anti-MiRNA Oligonucleotides (AMOs): Ammunition to Target MiRNAs Implicated in Human Disease?" *Gene Therapy*. 13 (6): 496-502. https://doi.org/10.1038/sj.gt.3302654.

Welty, Christopher J, Ilsa Coleman, Roger Coleman, Bryce Lakely, Jing Xia, Shu Chen, Roman Gulati, et al. 2013. "Single Cell Transcriptomic Analysis of Prostate Cancer Cells." *BMC Molecular Biology* 14 (1): 1-11. https://doi.org/10.1186/1471-2199-14-6.

Wilkerson, Matthew D, Jason M Schallheim, D Neil Hayes, Patrick J Roberts, Roy R. L. Bastien, Michael Mullins, Xiaoying Yin, et al. 2013. "Prediction of Lung Cancer Histological Types by RT-QPCR Gene Expression in FFPE Specimens." *Journal of Molecular Diagnostics* 15 (4): 485-97. https://doi.org/10.1016/j.jmoldx.2013.03.007.

Wu, Fengying, Jue Fan, Yayi He, Anwen Xiong, Jia Yu, Yixin Li, Yan Zhang, et al. 2021. "Single-Cell Profiling of Tumor Heterogeneity and the Microenvironment in Advanced Non-Small Cell Lung Cancer." *Nature Communications* 12 (1): 1-11. https://doi.org/10.1038/s41467-021-22801-0.

Wu, Hao, Xiaorong Li, and Hui Li. 2019. "Gene Fusions and Chimeric RNAs, and Their Implications in Cancer." *Genes and Diseases.* Elsevier 6 (4): *385-90.* https://doi.org/10.1016/j.gendis.2019.08.002.

Wu, Jia Qian, Jiang Du, Joel Rozowsky, Zhengdong Zhang, Alexander E Urban, Ghia Euskirchen, Sherman Weissman, Mark Gerstein, and Michael Snyder. 2008. "Systematic Analysis of Transcribed Loci in ENCODE Regions Using RACE Sequencing Reveals Extensive Transcription in the Human Genome." *Genome Biology* 9 (1): 1-14. https://doi.org/10.1186/gb-2008-9-1-r3.

Xi, Xiaochen, Tianxiao Li, Yiming Huang, Jiahui Sun, Yumin Zhu, Yang Yang, and Zhi John Lu. 2017. "RNA Biomarkers: Frontier of Precision Medicine for Cancer." *Non-Coding RNA.* 3(1): 9. https://doi.org/10.3390/ncrna3010009.

Xing, Shan, Xin Zheng, Li Qiang Wei, Shi Jian Song, Dan Liu, Ning Xue, Xiao Min Liu, et al. 2017. "Development and Validation of a Serum Biomarker Panel for the Detection of Esophageal Squamous Cell Carcinoma through RNA Transcriptome Sequencing." *Journal of Cancer* 8 (12): 2346-55. https://doi.org/10.7150/jca.19465.

Yang, Ningning. 2015. "An Overview of Viral and Nonviral Delivery Systems for MicroRNA." *International Journal of Pharmaceutical Investigation. Wolters Kluwer-Medknow Publications* 5 (4): 179. https://doi.org/10.4103/2230-973X.167646.

Yang, Yuchong, Xuanzhang Huang, Likun Zhou, Ting Deng, Tao Ning, Rui Liu, Le Zhang, et al. 2019. "Clinical Use of Tumor Biomarkers in Prediction for Prognosis and Chemotherapeutic Effect in Esophageal Squamous Cell Carcinoma." *BMC Cancer* 19 (1): 1-10. https://doi.org/10.1186/s12885-019-5755-5.

Yoshikatsu Koga, Masahiro Yasunaga, Amane Takahashi, Junichiro Kuroda, Yoshihiro Moriya, Takayuki Akasu, Shin Fujita, Seiichiro Yamamoto, Hideo Baba, Yasuhiro Matsumura. 2010 "MicroRNA Expression Profiling of Exfoliated Colonocytes Isolated from Feces for Colorectal Cancer Screening." *Cancer Prev Res (Phila)* 3 (11): 1435-42. https://doi.org/10.1158/1940-6207.CAPR-10-0036.

Yu, Bin, and Xuemei Chen. 2010. "Analysis of MiRNA Modifications." *Methods in Molecular Biology (Clifton, N.J.)* 592: 137-48. https://doi.org/10.1007/978-1-60327-005-2_10.

Yu, Kai, Yuqiong Hu, Fan Wu, Qiufang Guo, Zenghui Qian, Waner Hu, Jing Chen, et al. 2020. "Surveying Brain Tumor Heterogeneity by Single-Cell RNA-Sequencing of Multi-Sector Biopsies." *National Science Review* 7 (8): 1306-18. https://doi.org/10.1093/nsr/nwaa099.

Yu, Tonghu, Huaping Zhang, and Hong Qi. 2018. "Transcriptome profiling analysis reveals biomarkers in colon cancer samples of various differentiation." *Oncology Letters* 16(1): 48-54. https://doi.org/10.3892/ol.2018.8668.

Zhang, Yuping, Sathiya P Narayanan, Rahul Mannan, Gregory Raskind, Xiaoming Wang, Pankaj Vats, Fengyun Su, et al. 2021. "Single-Cell Analyses of Renal Cell Cancers Reveal Insights into Tumor Microenvironment, Cell of Origin, and Therapy Response." *Proceedings of the National Academy of Sciences of the United States of America* 118 (24). https://doi.org/10.1073/pnas.2103240118.

Zheng, Xin, Shan Xing, Xiao-Min Liu, Wen Liu, Dan Liu, Pei-Dong Chi, Hao Chen et al. "Establishment of using serum YKL-40 and SCCA in combination for the diagnosis of patients with esophageal squamous cell carcinoma." *Bmc Cancer* 14 (1) (2014): 1-10. https://doi.org/10.1186/1471-2407-14-490.

Zhou, Fei Fei, Rong Huang, Jun Jiang, Xiao Hong Zeng, and Shu Qian Zou. 2018. "Correlated Non-Nuclear COX2 and Low HER2 Expression Confers a Good Prognosis in Colorectal Cancer." *Saudi Journal of Gastroenterology* 24 (5): 301-6. https://doi.org/10.4103/sjg.SJG_46_18.

Zhou, Xiangxiang, Linquan Zhan, Kai Huang, and Xin Wang. 2020. "The Functions and Clinical Significance of CircRNAs in Hematological Malignancies." *Journal of Hematology and Oncology. Springer*. https://doi.org/10.1186/s13045-020-00976-1.

Zivicova, Veronika, Peter Gal, Alzbeta Mifkova, Stepan Novak, Herbert Kaltner, Michal Kolar, Hynek Strnad, et al. 2018. "Detection of Distinct Changes in Gene-Expression Profiles in Specimens of Tumors and Transition Zones of Tenascin-Positive/-Negative Head and Neck Squamous Cell Carcinoma." *Anticancer Research* 38 (3): 1279-90. https://doi.org/10.21873/anticanres.12350.

Zubovic, Lorena, Silvano Piazza, Toma Tebaldi, Luca Cozzuto, Giuliana Palazzo, Viktoryia Sidarovich, Veronica De Sanctis, et al. 2020. "The Altered Transcriptome of Pediatric Myelodysplastic Syndrome Revealed by RNA Sequencing." *Journal of Hematology and Oncology* 13 (1): 1-5. https://doi.org/10.1186/s13045-020-00974-3.

Chapter 4

Single-Cell Transcriptomics in Cancer Stem Cells (CSCs): Implications in Carcinogenesis and Drug Resistance

Sonam Sinha[1,*], Abhilasha Sharma[2] and Hamidullah Khan[3,†]

[1]Kashiv Biosciences, Block-B Opp. Apple Woods Township, Ahmedabad, Gujarat, India
[2]Department of Life Science, Gujarat University, Ahmedabad, Gujarat, India
[3]Department of Dermatology, University of Wisconsin-Madison, Madison, WI, USA

Abstract

Cancer cells are complex entities encompassing multiple cell types, resulting in molecular level heterogeneity leading to distinct and unequivocal clinical implications. Cancer stem cells (CSCs), a subset of specialized cells, play a critical role in the developing intratumoral complexities and functional diversity in these cells. The CSCs are described by self-renewal, multi-directional differentiation, inducing drug resistance, and carcinogenesis properties. Despite overall enhancement in cancer therapeutics, effectual therapy has been impeded by the manifestation of CSCs driven tumor cellular complexity that enterprise drug resistance and metastasis. Over the last decade, progress in medical sciences has steered towards the advancement of high throughput machineries such as single-cell transcriptomics (SCTs), which can analyse the gene expression and sequence report from individual cells of the minority population of cells including CSCs. SCTs provide the means to detect CSCs, study intratumoral heterogeneity, divulge the mechanism of drug resistance, and eventually direct personalized therapy approaches. Although high throughput technologies

*Corresponding Author's Email: sonam.sinha@kashivbio.com.
†Corresponding Author's Email: hamidullah@wisc.edu.

such as RNA sequencing provide a detailed understanding of cancer-related processes, these technologies are limited in that they represent "bulk" information from thousands of cancer cells. CSCs play critical roles and often control overall system behaviour, but they are not well reflected in "bulk" measurements, making them difficult to identify and quantify. Nonetheless, the absence of ubiquitously established CSC markers and characteristics has created disagreement among different reports. Identification and molecular characterization of CSCs, and the gene signalling pathways that regulate their behaviour and communications with other cell types, have substantial therapeutic potential to develop targeted therapies against cancer. Therefore, it becomes vital to use SCT technologies to comprehend the complexity of biological systems involved in cancer progression. In contrast to bulk transcriptomics, SCT technologies proffer a chance for the equitable identification and investigation of these CSC populations present in meagre quantities and a quiescent or latent condition and to develop specific cancer therapeutics. Finally, SCTs can identify minor but significant drug-resistant cancer stem cell populations in heterogeneous tumors, which can be utilized to select suitable treatments. However, SCTs have their competencies and inadequacies in terms of throughput, sensitivity, accuracy, and precision. In this chapter, we outline the role of SCTs in CSCs and offer our personal views about prospects that SCTs hold in the field. Singularizing tumor heterogeneity into noticeable morphological and phenotypic profiles continues to be a thought-provoking yet imperative goal in cancer treatment and management. Notably, the occurrence of SCTs in recent years has delivered an influential tool in resolving this impediment. Once employed in clinical samples amid tumor types, SCT technologies may prompt exponential growth and develop insights to explore the multitude of signalling pathways involved in carcinogenesis and contribute to developing strategies to combat drug resistance.

Keywords: single-cell transcriptomics, heterogeneity, cancer stem cells (CSCs), drug resistance, metastasis introduction

1. Introduction

Mammalian tumors are highly heterogeneous whose susceptibility towards anti-neoplastic therapy differs considerably among patients (Unger, Witte, and David 2015). Generally accepted clinical principle ascribes the histological category and stage of the tumor, corresponding to the tumor-node-metastasis (TNM) system, to differentiate between the therapy response (Bezabeh et al.

2005). However, the comprehension of oncology is constantly changing due to the noticeable advancement in genomics. Since the success of the Human Genome Project and expansion of next-generation sequencing (NGS) techniques, the assembly, and functions of the genome, as well as different mechanisms of gene regulations, have been comprehensively investigated in the oncology settings (Casamassimi et al. 2017; Wheeler and Wang 2013). This expansion of information on the genomic background of neoplastic transformation fetches novel insight into the molecular rationalization of the pathogenesis of particular neoplasms and their targeted therapy (López-Lázaro 2010; Janket et al. 2017). The division of molecular genetics that has been growing rapidly over recent years is transcriptomics (Lowe et al. 2017). A transcriptome is a group of all RNA molecules recorded from the genome in a particular cell, at a specific developmental phase, and under definite physiological or pathological settings (Jacquier 2009; Z. Wang, Gerstein, and Snyder 2009). It comprises protein-coding RNAs (pcRNAs), also known as messenger RNAs (mRNAs), and non-coding RNAs (ncRNAs). From these RNAs, every individual molecule imparts a diverse range of functions in the cell and responds differently to biological stimuli (Mallardo, Poltronieri, and D'Urso 2008; Jaeger, Doherty, and Ideker 2012; Iwata et al. 2019; Popov et al. 2019). The transcriptome profile of a cell could be figured as a snapshot of the short-term cell state. Hence, its exploration delivers knowledge about the functionality of genes and discloses facts on the genome plasticity, gene expression regulation, and alterations of singular transcripts (Cieślik and Chinnaiyan 2018). Therefore, transcriptome analysis is a beneficial method for the study of continuously altering neoplastic cells at a molecular level. Attained inferences contribute directly to the exhaustive advancement of precision oncology (Byron et al. 2016; Roy Chowdhury & Chinnaiyan, 2016; L. Wang et al. 2017).

Here in this chapter, our focus is on single-cell transcriptomics (SCT). SCT scrutinizes the gene expression level of singular cells in a specified population by concurrently determining the concentration of mRNA of thousands of genes. Analysis of this transcriptomic data resolves heterogeneous cell populations, reforms cellular evolving trajectories, and models transcriptional dynamics. All of which were concealed while measuring bulk transcriptome (Kalisky et al. 2018). As a result, using existing bulk cell analysis to detect a rare subpopulation of tumors could miss subgroups of residual specialized populations that are often rare and difficult to isolate. However, the SCT technique has an advantage over bulk measurements and is beneficial in detecting rare subpopulations in a given tumor population. Subsequently, the SCT technique analyses intratumoral heterogeneity, divulge the mechanism of drug resistance, and eventually directs personalized therapy approaches. Since there is plenty of evidence to establish the presence of rare populations, i.e., cancer stem cells (CSCs), in various human cancers, subgroups of which are resilient towards conventional and non-conventional cancer therapy and continue to grow during diminution. The chapter summarizes the current role and application of SCT in unravelling CSC-driven heterogeneity, its impact on cancer progression, and drug resistance, emphasizing its potential in personalized therapies.

2. Role of CSCs in Tumor Heterogeneity: Impact on Cancer Progression and Drug Resistance

Mammalian tumors encompass numerous cell populations in constant evolution, contributing to the intra-tumor heterogeneity. This complex level of intra-tumor heterogeneity is considered a key driving force of cancer propagation and drug resistance. Amid the variety of cell populations, particular cancer cells referred to CSCs are supposed to be responsible for producing a heterogeneous tumor lesion and causing tumor relapse and propagation and therapy resistance (Magee, Piskounova, and Morrison 2012). This event probably reflects intratumoral heterogeneity, with varied responses to therapy in distinctive tumor subpopulations. This process is linked to the tumor's hierarchical structure, in which only a few CSCs are capable of tumor development (Magee, Piskounova, and Morrison 2012). There is currently sufficient evidence suggesting the presence of rare CSCs in some cancers, whose subpopulations are resilient to cancer treatment and survive in remission. The CSC hypothesis and stochastic model were developed to understand cancer incidence and growth in a better way (Shackleton et al. 2009; W. Wang et al. 2014). CSC hypothesis clenches that every tumor subpopulation is fashioned by similar stem or progenitor cells and undergoes the same genetic and epigenetic processes, resulting in diverse cancer phenotypes.

In contrast, the stochastic model holds that every tumor subpopulation is derived from stem, progenitor, or differentiated cell types. Subsequently amassing adequate genetic and epigenetic alterations, CSCs progressively transform into a neoplastic cell. Hence, both models appear to elucidate the intra-tumoral heterogeneity (Shackleton et al. 2009; W. Wang et al. 2014).

Additionally, CSCs contribute to tumor relapse, owing to their stem cell-like attributes. However, their slow division rates make them a problematic therapeutic target. A growing body of evidence has reinstated the crucial role of stem-like cells in cancer by ascertaining that similar characteristics are shared by stem cells and cancer cells. One amongst them is cellular plasticity-the capability to switch and embrace different cell fates in response to biological stimuli and stressors (Thong et al. 2020). For instance, in epithelial-derived tumors, EMT and MET plasticity are vital for primary tumors to transition into mesenchymal tumors to spread, metastasize, and re-epithelialize at the secondary site (Tsai and Yang 2013).

Numerous molecular mechanisms occur regarding CSCs driven drug resistance, including augmented drug emanation, target alterations, DNA injury repair, stimulation of auxiliary cascade, and resistance to apoptosis (Holohan et al. 2013). Additionally, CSCs can self-renew and differentiate into the heterogeneous lineages of malignant cells in response to cancer therapeutics; CSCs driven heterogeneity considerably aids in therapy resistance (Swanton 2012). Also, CSCs induce cell cycle arrest that maintains their capacity to resist different cancer therapies. However, conventional cancer therapeutics target the rapidly dividing tumor cells to induce apoptosis and successfully abolish the bulk of these cells. A subset of residual CSCs survives and augment tumor relapse due to their slow division rate and higher invasive and therapy resistance capabilities (Kalisky et al. 2018). Understanding the landscapes of CSCs is significant to ascertain the basis for the new era of efficacious treatment of cancer patients. Therefore, to comprehend the biological phenomenon of cancer propagation, metastasis, tumor relapse, and drug resistance, fine characterization of CSCs driven heterogeneity is requisite. Recent technological developments centered on sequencing single cells pave the way for describing CSCs driven heterogeneity in cancer cells. In the analysis of functional heterogeneity, SCT significantly upsurges our molecular understanding of cancer propagation and therapy resistance. The application of SCT in understanding CSCs landscapes in different cancers and their usage in personalized therapy is detailed in this chapter's subsequent sections.

3. Application of Single-Cell Transcriptomics in CSCs

As mentioned above that the advent of SCT technology can distinguish and decode the CSCs driven intra-tumoral heterogeneity, distill molecular types, and offer new means to overcome CSCs-driven drug resistance (Patel et al. 2014). SCT methods empower unprejudiced characterization of the cellular and functional diversity of tumor tissues by identifying CSCs populations. SCT techniques are employed to identify CSCs that are allegedly scarce in numbers and exist in an inert or a dormant state, eventually strategizing targeted anti-tumor therapy. Recently, scRNA-seq investigations verified the manifestation of populations with stem-cell-like and drug-resilient attributes in brain and skin cancer (Tirosh et al. 2016). For instance, in oligodendrogliomas, scRNA-seq revealed that CSCs support tumor propagation and show indications of proliferation, while other tumor cell types do not.

In leukemia, a small number of CSCs were highly resistant to therapy and were probably accountable for tumor relapse when the treatment was discontinued (Giustacchini et al. 2017). Further, Veltan and his group displayed the potential of the SCT approach in characterizing CSCs in acute myeloid leukemia (AML). For instance, in AML, leukemic stem cells (LSCs) trigger mortality however are challenging to segregate owing to their low abundance and high likeness to healthy hematopoietic stem cells (HSCs). Velten and his group employed scRNA-seq based tactic and identified LSC-specific gene expression programs and the classification of differentiation blocks stimulated by leukemic mutations (Velten et al. 2021). A different group of researchers employed this method to investigate higher than two thousand CSCs from chronic myeloid leukemia (CML) patients through the entire tumor progression. They found the heterogeneity of CML-stem cells (SCs) with a discrete molecular mark that specifically continued during comprehensive anti-leukemic treatment (Giustacchini et al. 2017).

The heterogeneity in hepatocellular carcinoma (HCC) is responsible for the ineffectiveness of current curative therapy. Recently in 2019, Ho and his team demonstrated the practicability of the SCT approach to anatomize the intra-tumoral heterogeneity and investigate the transcriptomic landscape of rare CSCs in HCC. They found two distinctive cellular subpopulations established on the protein levels of epithelial cell adhesion molecules. These cells possess an amplified expression of multiple oncogenes with specific gene signatures indicating a novel stemness-associated cell subclone in HCC. Notably, SCT identified rare and significant CSCs which might direct better precision medicine (Ho et al. 2019).

Similarly, another group of researchers demonstrated that HCC-CSCs are phenotypically, functionally, and transcriptionally dissimilar at the single-cell level. And they possess distinct gene signatures within different CSC subpopulations (Zheng et al. 2018). Taken together, SCT revealed that diverse HCC-CSCs are self-reliantly related to HCC prognosis, affects intra-tumoral heterogeneity and tumor progression, which should be taken into consideration while strategizing anti-HCC-CSCs therapy.

Furthermore, in human colorectal cancer (CRC), Wang and co-researchers found that rare CSCs occur in an inert state and exhibits stem-like property and phylogenetic relationship toward cancer epithelial cells (EPCs). Additionally, CSCs harbor distinctive gene signatures, mainly *NOTUM, SMOC*, *BAMBI*, *PHLDA*, and *TNFRSF,* associated with CRC prognosis. Thus, by evaluating the diversity of CSCs and their association to EPCs, the SCT approach revealed the intratumoral heterogeneity and different subsets within CRC (H. Wang et al. 2021). Recently, a total of 20,938 cells from cervical cancer (CC) and adjacent normal tissues were examined by scRNA-seq (Li et al. 2021). The researchers and team identified CSCs subpopulations with distinctive signature genes and unique biological functions and displayed CC prognoses. The researchers established a relation between normal endothelial cells (NECs) and tumor-derived endothelial cells (TECs), successfully decoding the intra-tumoral heterogeneity of CC, paving the way for anti- CC therapy.

Chen and his team used scRNA-seq data and substantiated that the stem-like state found in breast cancers is bipotent, which can develop into both basal and liminal states. This bipotency is correlated with the clinical result and elevated risk of basal breast cancer (Chen et al. 2019). Wu, along with his team, employed RNA sequencing of individual cells using Fluidigm's Polaris system to investigate the divergence of SUM149, a hormone-receptor-negative breast cancer cell line (Wu et al. 2020). Interestingly, data revealed that CSCs of SUM149 displayed varied protein levels of hormone receptors, differential expressions of CD44v6 exons, and mRNA levels of epidermal growth factor receptors. On the one hand, high expression CD44v6 in CSCs was reported to promote tumor propagation. In contrast, elevated EGFR-211 and EGFR-201 in CSCs displayed less invasive and highly proliferative properties. Hence, the data established that heterogeneity of CSCs regulates their further growth and invasive capabilities. Furthermore, using scRNA-seq transcripts outlining and cluster examination, Emma and her team discovered the CSC-like subsets in MDA-MB-231 cells. The CSCs-like subgroup was found to have the dynamic capabilities of differentiating to other subpopulations of MDA-MB-231 cells. In the CSC-like subgroup, scRNA-seq gene expression analysis revealed 14 substantially amplified transcripts linked to stem cell features and patient survival information. (Ren et al. 2021).

Additionally, the SCT approach revealed that patients who are more prone to breast cancer recurrence displayed greater protein levels of CSC bio-markers than patients who are less prone to cancer reversion. This indicates the implication of employing breast CSC biomarkers for breast cancer prognosis (Tong et al. 2018).

Notably, high throughput computational analysis is required to obtain appropriate biological insights regarding cellular kinds and conditions that lead to cancer progression, tumorigenesis, and clinical consequences. Application of computational techniques in scRNA-seq analysis helps us identify similar cellular kinds and conditions shared among patients and ailments from different scRNA-seq databases and estimate the fractions of cell kinds in bulk gene expression reports. It also helps differentiate malignant cells from non-malignant cells by identifying marker and fusion genes, the difference in copy-number, and somatic alteration in scRNA-seq reports. Further, information about cell-cell communication from the receptors and ligands encoding genes can be inferred. With the help of trajectory extrapolation and RNA velocity study, transcriptions dynamics could also be easily characterized (Fan, Slowikowski, and Zhang 2020).

4. Role of Single-Cell Transcriptomics in Personalised Therapy

Personalized cancer treatments target altered cells or influence the tumor microenvironment using knowledge about the patient and cancer-specific biological markers. Identifying CSCs biomarkers that associate with drug response or possess a role in tumorigenesis and thus serve as therapeutic targets is critical in this process (Vargas and Harris 2016). However, CSCs heterogeneity among inter and intratumor within a single tumor creates a lot of chaos that needs to be dissected. This knowledge will almost certainly be crucial in implementing targeted treatment rationally. For instance, therapy failure might occur in cancer if a patient's tumor comprises various CSCs subpopulations and only some respond to treatment (usually therapy given based on bulk analysis) (Figure 1A). Irrespective of mammoth success attained by targeted therapies for cancer treatment, the risk of tumor relapse and therapeutic resistance still seems to be a mystery that is yet to be deciphered. Identifying how alterations in a single cell of a patient's tumor collectively drive cellular dysfunction may resolve the problem. scRNA-seq can be utilized to determine every clone contained by a cancerous mass and the concerning biological markers and equivalent drug, facilitating efficacious treatment (Figure 1B).

Recent advances in SCTs offer great potential to discriminate functionally healthy cells from cancer cells at different tumor stages, allowing a more precise prognosis and effective treatment (Levitin, Yuan, and Sims 2018). For instance, Longitudinal sketching of patient samples using scRNA-seq (or biological indicators found with it) can be utilized to track disease progression and decide the best moment to intervene, based on the advantage and costs of treatment (Figure 1C).

Analysing samples pre- and post-therapy can aid in identifying refractory groups, as well as biological markers and deciphering the root cause of resistance (Figure 1D). Because of its sensitivity, scRNA-seq might be utilized in a clinical environment to discover uncommon disease-associated cells (such as minimal residual disease) that would otherwise go undetected by bulk analysis (Figure 1E). Clinicians will use these emerging tools to evaluate the co-expression trend of possible drug targets in cellular mass as well as tumorous tissue to help them choose a personalized, targeted therapeutic regime for a patient. Single-cell expression profiling can unveil that if the target is widely expressed or is restricted to a small subset of rare cells. It also reveals if potential combination treatment targets are found in redundant pathways or separate subpopulations. However, this may seem relatively straightforward, but these intratumoral characteristics are hardly ever tested among the patients enrolled in clinical trials. SCT analysis could be an asset for determining drug response. It allows researchers to forecast the success of combination therapy that could target multiple cell types or pathways simultaneously. In sum, it is believed that SCT will profoundly affect the establishment and application of molecular enrollment standards for cancer clinical trials and the molecular assessment of therapeutic response.

5. Limitation of SCT in CSCs Driven Cancer

Apart from the significant advances of SCTs, few limitations need to be addressed. Applying SCTs to solid tumor samples is hampered by the requirement for complicated dissociation methods to get viable, personalized fresh CSCs. This limitation is particularly significant since many studies have raised concerns about possible transcriptional alterations resulting from tissue management during sample aspiration and handling (Tung et al. 2017). Although a few researchers have attempted to circumvent this restriction by using cell lines or organoids, they have been unable to replicate the intricate interactions between CSCs and the tumor surroundings (Grün et al. 2015). Furthermore, several or even serial samples from the same patient will be required to comprehend the tumor's molecular origin, which is challenging with solid cancers. Due to the small volume of recovered material, less-invasive biopsy practices like fine-needle aspiration (FNA) are not feasible for conventional genomic examination. Still, they provide the scope of hope for the use of SCTs in clinical science (Baslan and Hicks 2017).

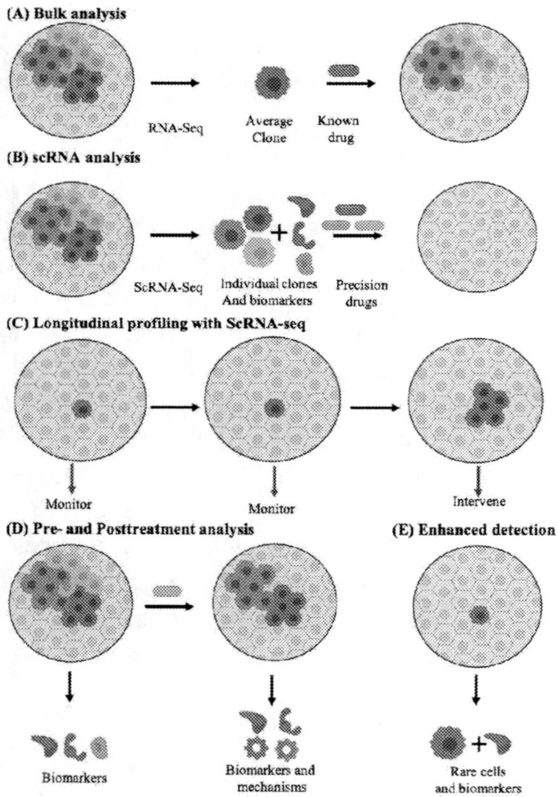

Figure 1. scRNA-seq applications in cancer medicine. (A) Bulk analysis of a tumor identifies the predominant malignant clone and suggests a drug to target it but not other clones. (B) scRNA-seq resolves each clone within a tumor and the corresponding biomarkers and cognate drugs, enabling successful therapy. (C) Longitudinal profiling of patient samples with scRNA-seq (D) Analysis of samples pre- and post-treatment may reveal drug-resistant biomarkers and mechanisms of resistance. (E) scRNA-seq in a clinical setting for detection of rare disease-associated cells.

Single nuclei RNA sequencing (sNuc-Seq) is an alternative technique to profile tumor gene expression from the nuclei. These nuclei are isolated from the frozen tissues instead of fresh tissues, critical to be segregated into a single-cell suspension for scrutiny (B. B. et al. 2016; N et al. 2016; R. V. et al. 2013; S. R. et al. 2016). Ruli Gao and coworkers manufactured a nanogrid and a microfluidic banking system that simultaneously facilitates imaging, selection, and sequencing of 100-1000s of individual cells or nuclei (Ruli et al. 2017). This enables us to assess if the nuclei are typical of the entire tumor cells. Another, limitation of SCTs is that it provides data exclusively about the transcriptomic state of a cell, but it comes at the expense of spatial information. Spatial information is vital to learning cellular functions more accurately. Therefore, to resolve this, recently, several multi-omic studies and analyses have been developed. One of them is the combination of cellular indexing of transcriptomes and epitopes by sequencing (CITE-seq) data with ultra-high multiplex immune-histochemistry (mIHC) data (Song et al. 2019; Hu et al. 2018).

Even though SCTs are currently in their early stages of use, they still have apparent benefits over traditionally existing transcriptomics approaches. Targeting increasingly varied transcriptomes and mapping the expression changes of uncommon isoforms across all genes is the next big challenge for SCTs. Pair-end sequencing, strand-specific sequencing, and the utilization of lengthier readouts to expand analysis and extent are all technologies that can help accomplish this aim.

Conclusion and Future Prospects

Transcriptomics unquestionably justifies being deliberated as a contemporary "game-changer" in tumor biology in molecular and clinical settings. Developments in individual-cell gene-expression approaches hold tremendous promise for understanding the CSCs diversity that underlies therapy resistance and provides the most significant challenge in precision cancer therapy. SCT allows an understanding of the molecular mechanisms triggering CSCs driven cancer, introducing trials associated with transcriptome-based personalized oncology. And hence, in the future, these SCT techniques might form the selection criteria of targeted amalgamation therapies and enrolment criteria for clinical trials. The SCT offers information related to molecular relationships between genetic information and the proteome, resulting in thorough insight into biological phenomena such as CSCs driven cancer. Still, compared to single-dimensional analysis, the multi-dimensional studies connect the transcriptome with other "omics" to methodically understand single-cell processes and behavior. Therefore, in the future, SCT analysis combined with multi-omic studies will add more insight into CSCs driven cancer. Collectively, SCT techniques are presented with new challenges. However, increased resolution of single-cell RNA sequencing information and enhanced statistical power of rapidly growing datasets will build the next steps for SCT to become a gold standard of modern cancer medicine.

References

Baslan, Timour, and James Hicks. 2017. "Unravelling Biology and Shifting Paradigms in Cancer with Single-Cell Sequencing." *Nature Reviews Cancer*. Nature Publishing Group. https://doi.org/10.1038/nrc.2017.58.

Bezabeh, Tedros, Olva Odlum, Richard Nason, Paul Kerr, Donna Sutherland, Rakesh Patel, and Ian C. P. Smith. 2005. "Prediction of Treatment Response in Head and Neck Cancer by Magnetic Resonance Spectroscopy." *American Journal of Neuroradiology* 26 (8): 2108–13. /pmc/articles/PMC8148844/.

Byron, Sara A., Kendall R. van Keuren-Jensen, David M. Engelthaler, John D. Carpten, and David W. Craig. 2016. "Translating RNA Sequencing into Clinical Diagnostics: Opportunities and Challenges." *Nature Reviews Genetics*. Nature Publishing Group. https://doi.org/10.1038/nrg.2016.10.

Casamassimi, Amelia, Antonio Federico, Monica Rienzo, Sabrina Esposito, and Alfredo Ciccodicola. 2017. "Transcriptome Profiling in Human Diseases: New Advances and Perspectives." *International Journal of Molecular Sciences*. MDPI AG. https://doi.org/10.3390/ijms18081652.

Chen, Weiyan, Samuel J. Morabito, Kai Kessenbrock, Tariq Enver, Kerstin B. Meyer, and Andrew E. Teschendorff. 2019. "Single-Cell Landscape in Mammary Epithelium Reveals Bipotent-like Cells Associated with Breast Cancer Risk and Outcome." *Communications Biology* 2 (1). https://doi.org/10.1038/s42003-019-0554-8.

Cieślik, Marcin, and Arul M. Chinnaiyan. 2018. "Cancer Transcriptome Profiling at the Juncture of Clinical Translation." *Nature Reviews Genetics*. Nature Publishing Group. https://doi.org/10.1038/nrg.2017.96.

Fan, Jean, Kamil Slowikowski, and Fan Zhang. 2020. "Single-Cell Transcriptomics in Cancer: Computational Challenges and Opportunities." *Experimental & Molecular Medicine 2020 52:9* 52 (9): 1452–65. https://doi.org/10.1038/s12276-020-0422-0.

Gao, R., Kim C., Sei E., Foukakis T., Crosetto N., Chan L. K., Srinivasan M., Zhang H., Meric-Bernstam F., and Navin N. 2017. "Nanogrid Single-Nucleus RNA Sequencing Reveals Phenotypic Diversity in Breast Cancer." *Nature Communications* 8 (1). https://doi.org/10.1038/S41467-017-00244-W.

Giustacchini, Alice, Supat Thongjuea, Nikolaos Barkas, Petter S. Woll, Benjamin J. Povinelli, Christopher A. G. Booth, Paul Sopp, et al. 2017. "Single-Cell Transcriptomics Uncovers Distinct Molecular Signatures of Stem Cells in Chronic Myeloid Leukemia." *Nature Medicine* 23 (6): 692–702. https://doi.org/10.1038/nm.4336.

Grindberg, R. V., Yee-Greenbaum J. L., McConnell M. J., Novotny M., O'Shaughnessy A. L., Lambert G. M., Araúzo-Bravo M. J., et al. 2013. "RNA-Sequencing from Single Nuclei." *Proceedings of the National Academy of Sciences of the United States of America* 110 (49): 19802–7. https://doi.org/10.1073/PNAS.1319700110.

Grün, Dominic, Anna Lyubimova, Lennart Kester, Kay Wiebrands, Onur Basak, Nobuo Sasaki, Hans Clevers, and Alexander van Oudenaarden. 2015. "Single-Cell Messenger RNA Sequencing Reveals Rare Intestinal Cell Types." *Nature* 525 (7568): 251–55. https://doi.org/10.1038/nature14966.

Habib, N., Li Y., Heidenreich M., Swiech L., Avraham-Davidi I., Trombetta J. J., Hession C., Zhang F., and Regev A. 2016. "Div-Seq: Single-Nucleus RNA-Seq Reveals Dynamics of Rare Adult Newborn Neurons." *Science (New York, N.Y.)* 353 (6302): 925–28. https://doi.org/10.1126/SCIENCE.AAD7038.

Ho, Daniel Wai Hung, Yu Man Tsui, Karen Man Fong Sze, Lo Kong Chan, Tan To Cheung, Eva Lee, Pak Chung Sham, Stephen Kwok Wing Tsui, Terence Kin Wah Lee, and Irene Oi Lin Ng. 2019. "Single-Cell Transcriptomics Reveals the Landscape of Intra-Tumoral Heterogeneity and Stemness-Related Subpopulations in Liver Cancer." *Cancer Letters* 459 (September): 176–85. https://doi.org/10.1016/j.canlet.2019.06.002.

Holohan, Caitriona, Sandra van Schaeybroeck, Daniel B. Longley, and Patrick G. Johnston. 2013. "Cancer Drug Resistance: An Evolving Paradigm." *Nature Reviews Cancer*. Nat Rev Cancer. https://doi.org/10.1038/nrc3599.

Hu, Youjin, Qin An, Katherine Sheu, Brandon Trejo, Shuxin Fan, and Ying Guo. 2018. "Single Cell Multi-Omics Technology: Methodology and Application." *Frontiers in Cell and Developmental Biology* 6 (APR). https://doi.org/10.3389/fcell.2018.00028.

Iwata, Michio, Longhao Yuan, Qibin Zhao, Yasuo Tabei, Francois Berenger, Ryusuke Sawada, Sayaka Akiyoshi, Momoko Hamano, and Yoshihiro Yamanishi. 2019. "Predicting Drug-Induced Transcriptome Responses of a Wide Range of Human Cell Lines by a Novel Tensor-Train Decomposition Algorithm." In *Bioinformatics*, 35:i191–99. Oxford University Press. https://doi.org/10.1093/bioinformatics/btz313.

Jacquier, Alain. 2009. "The Complex Eukaryotic Transcriptome: Unexpected Pervasive Transcription and Novel Small RNAs." *Nature Reviews Genetics*. Nat Rev Genet. https://doi.org/10.1038/nrg2683.

Jaeger, Philipp A., Colleen Doherty, and Trey Ideker. 2012. "Modeling Transcriptome Dynamics in a Complex World." *Cell*. Elsevier B.V. https://doi.org/10.1016/j.cell.2012.11.031.

Janket, Sok-Ja, Marium Qureshi, Antonio Bascones-Martinez, Jerian González-Febles, and Jukka H Meurman. 2017. "Holistic Paradigm in Carcinogenesis: Genetics, Epigenetics, Immunity, Inflammation and Oral Infections." *World Journal of Immunology* 7 (2): 11. https://doi.org/10.5411/wji.v7.i2.11.

Kalisky, Tomer, Sarit Oriel, Tali Hana Bar-Lev, Nissim Ben-Haim, Ariel Trink, Yishay Wineberg, Itamar Kanter, Shlomit Gilad, and Saumyadipta Pyne. 2018. "A Brief Review of Single-Cell Transcriptomic Technologies." *Briefings in Functional Genomics*. Oxford University Press. https://doi.org/10.1093/bfgp/elx019.

Krishnaswami, S. R., Grindberg R. V., Novotny M., Venepally P., Lacar B., Bhutani K., Linker S. B., et al. 2016. "Using Single Nuclei for RNA-Seq to Capture the Transcriptome of Postmortem Neurons." *Nature Protocols* 11 (3): 499–524. https://doi.org/10.1038/NPROT.2016.015.

Lake, B. B., Ai R., Kaeser G. E., Salathia NS, Yung Y. C., Liu R., Wildberg A., et al. 2016. "Neuronal Subtypes and Diversity Revealed by Single-Nucleus RNA Sequencing of the Human Brain." *Science* (New York, N.Y.) 352 (6293): 1586–90. https://doi.org/10.1126/SCIENCE.AAF1204.

Levitin, Hanna Mendes, Jinzhou Yuan, and Peter A. Sims. 2018. "Single-Cell Transcriptomic Analysis of Tumor Heterogeneity." *Trends in Cancer*. Cell Press. https://doi.org/10.1016/j.trecan.2018.02.003.

Li, Chunbo, Luopei Guo, Shengli Li, and Keqin Hua. 2021. "Single-Cell Transcriptomics Reveals the Landscape of Intra-Tumoral Heterogeneity and Transcriptional Activities of ECs in CC." *Molecular Therapy - Nucleic Acids* 24 (June): 682–94. https://doi.org/10.1016/J.OMTN.2021.03.017.

López-Lázaro, Miguel. 2010. "A New View of Carcinogenesis and an Alternative Approach to Cancer Therapy." *Molecular Medicine*. The Feinstein Institute for Medical Research. https://doi.org/10.2119/molmed.2009.00162.

Lowe, Rohan, Neil Shirley, Mark Bleackley, Stephen Dolan, and Thomas Shafee. 2017. "Transcriptomics Technologies." *PLoS Computational Biology* 13 (5). https://doi.org/10.1371/journal.pcbi.1005457.

Magee, Jeffrey A., Elena Piskounova, and Sean J. Morrison. 2012. "Cancer Stem Cells: Impact, Heterogeneity, and Uncertainty." *Cancer Cell*. Cancer Cell. https://doi.org/10.1016/j.ccr.2012.03.003.

Mallardo, Massimo, Palmiro Poltronieri, and Oscar Fernando D'Urso. 2008. "Non-Protein Coding RNA Biomarkers and Differential Expression in Cancers: A Review." *Journal of Experimental and Clinical Cancer Research*. BioMed Central. https://doi.org/10.1186/1756-9966-27-19.

Patel, Anoop P., Itay Tirosh, John J. Trombetta, Alex K. Shalek, Shawn M. Gillespie, Hiroaki Wakimoto, Daniel P. Cahill, et al. 2014. "Single-Cell RNA-Seq Highlights Intratumoral Heterogeneity in Primary Glioblastoma." *Science* 344 (6190): 1396–1401. https://doi.org/10.1126/science.1254257.

Popov, X. Daniil V., Pavel A. Makhnovskii, Elena I. Shagimardanova, Guzel R. Gazizova, Evgeny A. Lysenko, Oleg A. Gusev, and Olga L. Vinogradova. 2019. "Contractile Activity-Specific Transcriptome Response to Acute Endurance Exercise and Training in Human Skeletal Muscle." *American Journal of Physiology - Endocrinology and Metabolism* 316 (4): E605–14. https://doi.org/10.1152/ajpendo.00449.2018.

Ren, Lili, Junyi Li, Chuhan Wang, Zheqi Lou, Shuangshu Gao, Lingyu Zhao, Shuoshuo Wang, et al. 2021. "Single Cell RNA Sequencing for Breast Cancer: Present and Future." *Cell Death Discovery* 7 (1): 104. https://doi.org/10.1038/s41420-021-00485-1.

Roychowdhury, Sameek, and Arul M. Chinnaiyan. 2016. "Translating Cancer Genomes and Transcriptomes for Precision Oncology." *CA: A Cancer Journal for Clinicians* 66 (1): 75–88. https://doi.org/10.3322/caac.21329.

Shackleton, Mark, Elsa Quintana, Eric R. Fearon, and Sean J. Morrison. 2009. "Heterogeneity in Cancer: Cancer Stem Cells versus Clonal Evolution." *Cell*. Elsevier B.V. https://doi.org/10.1016/j.cell.2009.08.017.

Song, Yanling, Xing Xu, Wei Wang, Tian Tian, Zhi Zhu, and Chaoyong Yang. 2019. "Single Cell Transcriptomics: Moving towards Multi-Omics." *Analyst* 144 (10): 3172–89. https://doi.org/10.1039/c8an01852a.

Swanton, Charles. 2012. "Intratumor Heterogeneity: Evolution through Space and Time." *Cancer Research*. Cancer Res. https://doi.org/10.1158/0008-5472.CAN-12-2217.

Thong, Tasha, Yutong Wang, Michael D. Brooks, Christopher T. Lee, Clayton Scott, Laura Balzano, Max S. Wicha, and Justin A. Colacino. 2020. "Hybrid Stem Cell States: Insights Into the Relationship Between Mammary Development and Breast Cancer Using Single-Cell Transcriptomics." *Frontiers in Cell and Developmental Biology* 8 (May). https://doi.org/10.3389/fcell.2020.00288.

Tirosh, Itay, Andrew S. Venteicher, Christine Hebert, Leah E. Escalante, Anoop P. Patel, Keren Yizhak, Jonathan M. Fisher, et al. 2016. "Single-Cell RNA-Seq Supports a Developmental Hierarchy in Human Oligodendroglioma." *Nature* 539 (7628): 309–13. https://doi.org/10.1038/nature20123.

Tong, Mengying, Ziqian Deng, Mengying Yang, Chang Xu, Xiaolong Zhang, Qingzheng Zhang, Yuwei Liao, et al. 2018. "Transcriptomic but Not Genomic Variability Confers Phenotype of Breast Cancer Stem Cells 06 Biological Sciences 0604 Genetics 11 Medical and Health Sciences 1112 Oncology and Carcinogenesis." *Cancer Communications* 38 (1). https://doi.org/10.1186/s40880-018-0326-8.

Tsai, Jeff H., and Jing Yang. 2013. "Epithelial-Mesenchymal Plasticity in Carcinoma Metastasis." *Genes and Development*. Cold Spring Harbor Laboratory Press. https://doi.org/10.1101/gad.225334.113.

Tung, Po Yuan, John D. Blischak, Chiaowen Joyce Hsiao, David A. Knowles, Jonathan E. Burnett, Jonathan K. Pritchard, and Yoav Gilad. 2017. "Batch Effects and the Effective Design of Single-Cell Gene Expression Studies." *Scientific Reports* 7 (January). https://doi.org/10.1038/srep39921.

Unger, Florian T., Irene Witte, and Kerstin A. David. 2015. "Prediction of Individual Response to Anticancer Therapy: Historical and Future Perspectives." *Cellular and Molecular Life Sciences*. Birkhauser Verlag AG. https://doi.org/10.1007/s00018-014-1772-3.

Vargas, Ashley J., and Curtis C. Harris. 2016. "Biomarker Development in the Precision Medicine Era: Lung Cancer as a Case Study." *Nature Reviews Cancer*. Nature Publishing Group. https://doi.org/10.1038/nrc.2016.56.

Velten, Lars, Benjamin A. Story, Pablo Hernández-Malmierca, Simon Raffel, Daniel R. Leonce, Jennifer Milbank, Malte Paulsen, et al. 2021. "Identification of Leukemic and Pre-Leukemic Stem Cells by Clonal Tracking from Single-Cell Transcriptomics." *Nature Communications* 12 (1). https://doi.org/10.1038/s41467-021-21650-1.

Wang, Hua, Peng Gong, Tong Chen, Shan Gao, Zhenfeng Wu, Xiaodong Wang, Jie Li, et al. 2021. "Colorectal Cancer Stem Cell States Uncovered by Simultaneous Single-Cell Analysis of Transcriptome and Telomeres." *Advanced Science* 8 (8). https://doi.org/10.1002/advs.202004320.

Wang, Lu, Lipi Acharya, Changxin Bai, and Dongxiao Zhu. 2017. "Transcriptome Assembly Strategies for Precision Medicine." *Quantitative Biology*. Higher Education Press. https://doi.org/10.1007/s40484-017-0109-2.

Wang, Weikang, Yi Quan, Qibin Fu, Yu Liu, Ying Liang, Jingwen Wu, Gen Yang, Chunxiong Luo, Qi Ouyang, and Yugang Wang. 2014. "Dynamics between Cancer Cell Subpopulations Reveals a Model Coordinating with Both Hierarchical and Stochastic Concepts." *PLoS ONE* 9 (1). https://doi.org/10.1371/journal.pone.0084654.

Wang, Zhong, Mark Gerstein, and Michael Snyder. 2009. "RNA-Seq: A Revolutionary Tool for Transcriptomics." *Nature Reviews Genetics*. Nat Rev Genet. https://doi.org/10.1038/nrg2484.

Wheeler, David A., and Linghua Wang. 2013. "From Human Genome to Cancer Genome: The First Decade." *Genome Research*. Genome Res. https://doi.org/10.1101/gr.157602.113.

Wu, Shaocheng, Hongjiu Zhang, Shamileh Fouladdel, Hongyang Li, Evan Keller, Max S. Wicha, Gilbert S. Omenn, Ebrahim Azizi, and Yuanfang Guan. 2020. "Cellular, Transcriptomic and Isoform Heterogeneity of Breast Cancer Cell Line Revealed by Full-Length Single-Cell RNA Sequencing." *Computational and Structural Biotechnology Journal* 18 (January): 676–85. https://doi.org/10.1016/j.csbj.2020.03.005.

Zheng, Hongping, Yotsawat Pomyen, Maria Olga Hernandez, Caiyi Li, Ferenc Livak, Wei Tang, Hien Dang, et al. 2018. "Single-Cell Analysis Reveals Cancer Stem Cell Heterogeneity in Hepatocellular Carcinoma." *Hepatology* 68 (1): 127–40. https://doi.org/10.1002/hep.29778.

Chapter 5

Transcriptome: A Tool for Precision Oncology

Lalita Dwivedi[1,*] and Km Anjaly[2]

[1]Biological Product division, Indian veterinary Research Institute, Bareilly, India

[2]School of Life Sciences, Jawaharlal Nehru University, New Delhi, India

Abstract

In the evolving technological world, we have made tremendous progress in biology and stepped past the 'omics era'. Omics refers to the collective analysis of large/meta data sets of biological variables comprising genes (genomics), RNAs (transcriptomics), protein (proteomics), lipids (lipidomics), epigenetics (epigenomics) and metabolites (metabolomics). One of the applications of omics includes the development of precision medicine which focuses on delivering personalized medication depending on the patient's need. Precision oncology aims at identifying the molecular markers of tumours to find out the therapeutics. Cancer is a disease of accumulating genetic and epigenetic changes in numerous genes, some of which drives tumour promotion and progression. These changes in the tumour microenvironment contribute to the phenotypic and functional heterogeneity within the same tumour making it resistant to the therapy. Transcriptomics comprises the expression of all the coding and non-coding RNA transcripts in an individual or population of cells providing a complete view of the expression dynamics. Transcriptomic studies give an insight into the gene expression variability in the tumours and their microenvironment. Different types of transcriptome technologies such as RNA sequencing, microarray and single-cell transcriptomics have evolved with time, and help us to design cancer precision medicine by identifying the actionable biomarkers. These techniques also act as good predictors of the response to the therapies providing an edge to both the oncologists and patients. Moreover, these techniques help to identify disease-specific biomarkers which facilitate diagnosis. These may also be used to classify cancer into clinically relevant molecular subtypes based on the variability in the

---

*Corresponding Author's Email: lalitasharma19@gmail.com.

expression profile of tumour cells. In future, increased knowledge in the field of precision medicine can lead to early diagnosis and better prognosis compared to other traditional approaches such as histopathology. Recent development in the field of cancer transcriptome has revealed extensive studies using RNA sequencing and generated databases for various cancer cell lines and tissue samples. These databases have greatly aided in the identification of combination of target-biomarkers, aiding to the development of precision medicine based on the different types and stages of cancer. Database like TCGA, CCLE, NCI60, GDSC, LINCS and Project Achilles have the compilation of data for multiple cancer types obtained from various cancer patients and cell lines through RNA sequencing and DNA microarray assay. These databases have the record of a number of mutations and expression profiles of various genes. Thus, the data from these databases can be used for the design of precision medicine based on the targetable mutations and gene expression patterns. Transcriptome engineering to manipulate the loss of function and gain of function changes in the gene expression at transcriptome level is of high significance in the development of specifically designed drugs for cancer treatment. This kind of engineering utilizes different transcriptome engineering tools such as siRNAs, shRNAs and CRISPR-Cas system. This chapter describes the recent developments in the field of cancer transcriptomics that can be translated into clinical oncology.

Keywords: precision oncology, transcriptome, cancer, RNA sequencing

1. Introduction

Transcriptome represents the status of all the transcripts in a cell or tissue and gives an idea about transcriptionally active genes. It helps in determining the turned-on or off conditions of the genes. In this way, transcriptome profiling has evolved as one of the most commonly used tools for determining the root cause of any disease at the molecular level as it can assess alternate splicing, polyadenylation and can identify fusion transcripts (Casamassimi et al. 2017). Cancer is the second most common disease after cardiovascular disease worldwide. It is a dynamic disease that can affect any part of the human body. According to Globocan in 2018, 9.6 million deaths were reported due to cancer (Sung et al. 2021). Cancers usually become more heterogeneous over time and harbour distinct molecular signatures. This heterogeneity leads to the irregular distribution of genetically distinct cell sub-populations across the tumour sites. This difference in the molecular makeup within the same tumour site is the main hurdle for cancer treatment and the root cause of resistance (Meacham and Morrison 2013). Transcriptome profiling is a powerful method through which the assessment of tumour heterogeneity can be accomplished by identifying the molecules that are specific to tumours or for their subtypes. These molecular biomarkers can be helpful in the early detection, diagnosis, prognosis, and in deciding the therapeutic strategy.

Although alterations at the genetic level constitute the major focus of tumour-specific changes, but in many respects, mRNA holds a more promising approach due to the following reasons such as the copies of mRNA exist in thousand numbers allowing a highly sensitive and early detection/diagnosis. The immediate downstream effect of the genetic variations and drug treatment reflects directly into the transcriptome patterns. Thus, the transcriptome-based markers sensitively represent different cancer-related cellular changes. Furthermore, during transcriptome profiling, there is no need to consider various cancer mutation-associated aspects such as zygosity, LOF/GOF considerations and the position effect which makes genomic approaches more complex ((Jeong et al. 2017). It also reveals the potential candidate antigens for monoclonal antibody targeting, vaccine development, and the adoptive immunotherapies comprehensively and with high sensitivity (Adamia et al. 2014; Rousseaux et al. 2013). The efficacy of using mRNA for these purposes is highly dependent on the degree of tumour-specific expression. Several studies have also compared the metastasized tumour with the primary tumour to delineate the key molecular drivers which lead to the development of tumour cell invasion and metastasis.

In the last few years, personalized medicine is experiencing accelerated and exponential growth in the field of cancer therapeutics. Due to the inter-individual tumour heterogeneity, the existing cancer drugs are effective only in some of the patients while being ineffective in others. Here comes the concept of precision or personalized oncology where anti-cancer medicines are given according to the molecular profile of the cancer patients depending on the stage of cancer, thus delivering the right medicine to the right person at the right time. Precision oncology helps in maximizing the patient's survival and improving the life quality. Testing right is the first step in the application of precision oncology. Thus, the search for suitable biomarkers is a mandate for precision oncology. Transcriptomic data of patients obtained by RNA sequencing or other methods and transcriptome-based assays utilizing si/shRNA and CRISPRs help in the identification of suitable targets for cancer treatment at the transcriptional level, therefore aiding in the design of precision medicines for cancer treatment (Jeong et al. 2017).

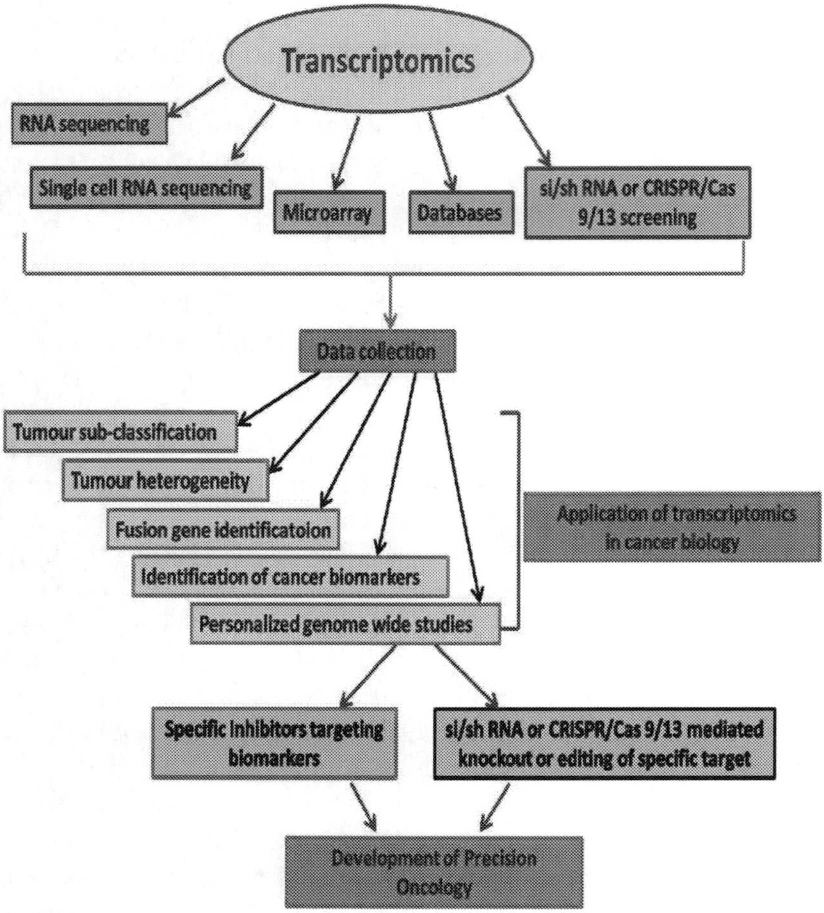

Figure 1. Transcriptomic tools in the precision oncology.

For cancer diagnosis, the transcriptomic tools have an advantage over the histological examinations as they do not require invasive retrieval of tissue, as well as they can clinically distinguish the relevant subtypes of cancer which is not possible in the histological examinations. At present, commercial gene expression signature diagnostic tests are within the easy reach for various cancers but their use in the clinics is hindered by FDA regulation, biomarker selection, cost and insurance coverage. Genomic Health's Oncotype DX, Agendia's MammaPrint Breast Cancer Recurrence Test, Sividon Diagnostics' Endopredict, and Nanostring Technologies are some of the names of the available assays which provide information regarding the expression levels of relevant genes and helps in determining better treatment decisions for the cancer patients. The WINTHER is the first international clinical trial proposed by WIN Consortium to identify the predictive biomarkers beyond the genomic aberrations (Rodon et al. 2019).

2. Transcriptomics (Single Cell Transcriptomics (ScRNA)-Seq) in Delineating the TumourMicroenvironmentHeterogeneity

The single-cell transcriptomics or scRNA-seqapproach is used to identify the expression level of various genes at the individual cell level. Tang et al. firstly introduced the concept of scRNA-sequencing (Tang et al. 2009). Afterwards, several other protocols for scRNA-seq were developed such as DroNC-seq, SPLiT-seq (Rosenberg et al. 2018), CEL-seq2 (Hashimshony et al. 2016), MARS-seq (Jaitin et al. 2014), Smart-seq (Ramskold et al. 2012), Drop-seq (Ziegenhain et al. 2017), and MATQ-seq (Sheng et al. 2017). Single cell RNA sequencing is being widely used for studying the tumour microenvironment (TME) heterogeneity.

The tumour microenvironment (TME) is an important factor for tumour heterogeneity, as well as it has a significant contribution in tumour initiation, progression, and treatment response. TME consists of a diverse set of cell types such as cancer-associated fibroblasts (CAFs), infiltrating immune cells (IICs) and endothelial vascular cells (Wang et al. 2017). The cancer-associated fibroblasts (CAF) are key part of tumour microenvironment. Seven hundred and sixty eight CAFs isolated from a genetically modified mouse model of breast cancer (MMTV-PyMT) were analysed by single-cell RNA-sequencing (scRNA-seq) which delineates three transcriptionally diverse sub-populations of CAFs named as cycling CAFs (cCAFs), developmental CAFs (dCAFs) and matrix CAFs (mCAFs) causing heterogeneity within the tumour microenvironment. These three subtypes of CAFs co-exist within the TME and possess unique gene signature indicating their specialized function during cancer growth. The gene profile is functionally distinct for a particular CAF subtype, thereby it holds an independent prognostic value in the clinical outcome as well as it helps in the development of precision medicine (Bartoscheck et al. 2018).

IICs and different subtypes of immune cells communicate with the cancer cells and determine the balance between immunotolerance and immune-rejection. For making immunotherapy more effective, elimination of the immunosuppressive environment that is generated due to a high degree of heterogeneity in IICs would be needed. Such immunotherapeutic strategies are now a days available in the clinics, which directly and indirectly target immunosuppressive cells. Quantitative analysis of different tumour-IICs can serve as cancer biomarker and can predict patient survival as well as immunogenic effects of anticancer therapies, ultimately guiding the rational design of combination therapies. Most importantly, provided that the immunotherapy utilizing the immune-checkpoint blockers is effective only in a limited fraction of patients, thus combination therapies are highly preferable (Topalian et al. 2009).

Angiogenesis performs a vital role in the tumour growth and metastasis as new vessels are formed from pre-existing ones (Muthukkaruppan,Kubai, and Auerbach 1982). In the tumour microenvironment, the endothelial cells are exposed to the high levels of VEGF-A, and this deregulated signalling makes them unstable and causes polarity loss. This ultimately induces the establishment of endothelial podosomes which degrade the extracellular matrix composed of laminins, collagen and proteoglycans promoting the tumour endothelial cell invasiveness (Nagl et al. 2020). The role of tumour endothelial cells (TECs) in the tumour microenvironment is multifaceted. Along with promoting the tumour angiogenesis, they are also involved in the regulation of immune response as they can activate or downregulate the effector immune cells. They are also involved in the endothelial anergy, thereby directly affecting the immunotherapeutic responses suggesting that a better understanding of the functional traits of TECs is critical.

TECs are highly metabolically active cells and contain a fourfold increase in the RNA content, thus the transcriptome profile of TECs is markedly different from that of NECs (normal endothelial cells), and thereby the identification of possible TEC-unique genetic signatures is of great interest. ECs of human as well as mouse malignant tumours exhibit abnormal gene expression profiles and chromosomal instability (Akino et al. 2009; Schaaf, Garg, and Agostinis 2018). Using scRNAseq technology, several studies showed the heterogeneous origin of TECs (Lambrechts et al. 2018; Goveia et al. 2020; Rohlenova et al. 2020). Due to their stem cell-like origin, TECs can orchestrate tumour neo-angiogenesis. Moreover, Myc-mediated transcription is highly activated in TECs, which associates with the tumour-aggressiveness and is known to have a strong impact on the angiogenesis. Luo et al. in 2013 isolated the endothelial cells from the normal as well as the malignant prostate surgical specimens via CD31+ marker selection method and found an enormous difference in the transcriptome and methylome between prostate tumour-derived endothelial cells (TdECs) and the normal-tissue derived endothelial cells (NdECs) and thus providing new anti-angiogenic targets

(Luo et al. 2013). Fifty six thousand seven hundred and seventy one endothelial cells (ECs) isolated from different sources including cultured lung TECs and lung tumoral tissue of human and mouse were sequenced, which revealed the endothelial cell heterogeneity of lung cancer and the therapeutically targetable angiogenic candidates. Along with this, 17 known and 16 previously unrecognized phenotypes were also found (Goveia et al. 2020).

3. Transcriptomics in the Sub-Classification of Cancers According to the Clinical Relevance

Transcriptomics has also aided in the classification of cancers based on the clinical relevance. For example, classification of B cell lymphoma in the activated B cell lymphoma and germinal centre B cell lymphoma was done on the basis of transcriptomics data of which theactivated B-cell lymphoma have poor prognosis as compared to the germinal centre lymphoma) (Alizadeh et al. 2000; Barton et al. 2012). Biologically and clinically relevant five subsets of breast cancer were identified by Perou et al. based on the transcriptomic profiling (Perou et al. 2000). Ellis et al. stratified lymphoma and breast cancer into clinically significant subtypes on the basis of the transcriptome as well as genome profiling (Ellis and Perou 2013). Apart from the sub-classification of tumours, the mRNA and miRNA expression profile based assays can be used to identify the tissue of origin of cancers with unknown primary origin (Hainsworth et al. 2013; Varadhachary et al. 2011).

4. Transcriptomics in the Identification of Cancer-Related Fusion Genes

Gene fusions can be detected at the transcript level by the way of paired-end sequencing method (Maher, Kumar-Sinha, et al. 2009; Maher, Palanisamy, et al. 2009). Stransky et al. comprehensively analyzed RNA sequencing data belonging to 7000 cancer samples from the TCGA database for the detection of gene fusion events. In this study, they were able to find several known and novel gene fusions including kinases (Stransky et al. 2014). In a similar study, involving 675 cancer cell lines, 2200 gene fusions were catalogued, out of which, 1435 fusions included novel fusion partner genes which were not previously reported to take part in the fusion events (Klijn et al. 2015). Roberts et al. reported various kinase fusions in B-cell lymphoblastic leukemia of children including known genes for which targeted therapies are available such as Janus kinase 2(JAK2), a neurotrophic tyrosine kinase receptor 3 (NTRK3), Abelson Murine Leukemia Viral Oncogene Homolog 1 (ABL1) and colony-stimulating factor 1 receptor (CSFR1), thus providing new therapeutic approach for the patients with this type of leukemia (Kathryn G. Roberts et al. 2012;K.G. Roberts et al. 2014). Similarly, targeting the ALK gene in the metastatic lung cancer patients having gene fusions involving the ALK kinase represents a biomarker for clinical outcome (Shaw and Engelman 2014; Shaw et al. 2013). Thus, such gene fusion studies can be useful in designing the personalized medicines for the patients having gene fusions by targeting the genes involved in the fusion events, for example, kinase inhibitors can be tested for this purpose in the clinical trials.

5. Transcriptomics in the Identification of Cancer Biomarkers in Different Types of Cancers

Transcriptome analysis also has great potential in the identification of various diagnostic, prognostic and predictive biomarkers for delineating the cancer type and subtype (diagnosis), for the determination of disease progression or relapse after therapy (prognosis), and the identification of molecular targets for the targeted drug design (predictive biomarker). For example, the OncotypeDx test has a prognostic as well as a predictive function as it helps in deciding whether the patient needs any additional therapy after surgery representing its prognostic potential and also help in deciding the patients who can get benefitted from the adjuvant therapy representing its function as a predictive marker. This test performs the assessment of 21 transcripts in the females having ER^+ and lymph node negative breast cancer (Paik et al. 2004). RNA expression analysis of the Epstein-Barr virus makes a standard diagnostic marker for lymphoma (Roychowdhury and Chinnaiyan 2016). ALK gene fusion acts as a predictive biomarker in the advanced lung cancer for the assessment of clinical significance of using ALK inhibitors (Shaw and Engelman 2014; Shaw et al. 2013).

Breast cancer exhibits variability in the molecular profile of the individual cells, indicating its heterogeneity. These molecular variabilities classify breast cancer into different subtypes. These subtypes respond differently to the therapy leading to differential sensitivity of chemotherapy and other therapies to different cancer subtypes. $HER2^{+}$, luminal-like, basal-like, and normal like breast cancer are the four subtypes of breast cancer. Hormone therapy and anti-HER2 therapies cannot be applied to triple-negative breast cancer (TNBC) {Estrogen receptor (ER) (-), progesterone receptor (PR) (-) and human epidermal growth factor receptor 2 (HER2) (-)}, and the efficacy of available therapy is still unclear. Along with this, TNBC also lacks effective prognostic and therapeutic markers. RNA-seq data of 165 TNBC samples and normal samples from both the surgically resected samples as well as the TCGA cohort identified 148 TNBC specific genes. Comparison of RNA-seq data of TNBC with the expression profiles of the adjacent normal samples and other subtypes revealed that these TNBC specific genes are associated with the hormone-related pathways and mammary gland morphogenesis, indicating the role of these pathways in TNBC and in differentiating TNBC from other subtypes. FSIP1, ATP1A2, FSD1, CMTM5, AFF3, ADCY5, CYP2A7, HMSD, C11orf86 are the nine genes that were associated with TNBC prognosis. Out of these genes, CYP2A7, ATP1A2 and ADCY5 are the three genes related to the hormone pathways that may be helpful in both the prognosis as well as the targeted therapy (Zhang et al. 2018).

After lung cancer, prostate cancer (PCa) is the second leading cause of cancer-related morbidity among men. Forty percent of the diagnosed cases of PCa are asymptomatic and the survival rate of localized PCa is 99% but it reduces to 30% once it metastasizes. Blood PSA determination is an invasive diagnostic procedure for PCa screening, but most of the time, false positive results leading to the over treatment have raised much concern in identifying other biomarkers with high accuracy. The next-generation sequencing (NGS) data of urinary samples from benign hyperplasia (BPH), high-stage and low-stage PCa patients screened various differentially expressed transcripts in the pair-wise comparison, mostly associated with the RAN signaling pathways and protein ubiquitination process. Polyhomeotic-like protein 3 (PHC3), bromodomain and PHD finger-containing protein 1 (BRPF1), ferritin heavy chain 1 (FTH1), uveal auto-antigen with coiled-coil domains, ankyrin repeats (UACA) and, oxysterol-binding protein 1 (OSBP) are the validated genes. A cohort study also validates that these biomarkers have the capability to discriminate high, intermediate and low risk patients which may be helpful in making the therapeutic decisions (Wang and Thorgeirsson 2009).

Hepatocellular carcinoma (HCC) diagnosis and prognosis is a difficult task due to the tumour stage and the compromised liver function because of viral hepatitis, hindering it from providing an accurate clinical recommendation. There is always a greatest risk for the hepatocellular carcinoma recurrence even after the patients have undergone potentially curative treatment for it. To identify such high-risk patients is a challenge. Wang et al. have profiled the tumoral as well as non-tumoral tissues from 307 HCC patients and screened the feasibility of more than 6000 human genes for the prediction of patient survival leading to the identification of a gene signature comprising of 186 genes from non-tumoral region showing correlation with the patient survival. It has been observed that the gene-expression profiles of the tumoral tissue has poor association with the patient's survival, whereas the gene-expression profiles of the surrounding non-tumoral liver tissue were highly associated with the survival (Wang and Thorgeirsson 2009).

Clear cell carcinoma (CCC) of both the ovarian and the uterine cancer exhibits a unique clinical behaviour and a histologic subtype. RNA sequencing data of ovarian CCCs and uterine CCCs were compared with publicly available data of high-grade serous ovarian cancers (HGSOCs) and both showed significant gene expression differences. Ovarian CCCs and uterine CCCs contain distinctive transcriptional expression profiles which distinguish them from each other and also distinguish the clear cell uterine cancer from the other more common histological subtypes which can be helpful in devising the novel therapeutics. Differences in the ARID1A gene expression and ethnicity categorize ovarian CCCs into two subgroups. Upregulation of 1,607 genes and downregulation of 109 genes were significantly observed in ovarian CCCs and uterine CCCs. Specific metabolic and PTEN pathways were predominated in ovarian CCCs. The programmed death-ligand1 helps in the immune evasion as well as is associated with the poor prognosis of ovarian and uterine cancer. Memory B-cell and resting NK cell populations were abundantly present in PD-L1^{+} ovarian CCCs, whereas PD-L1^{-} ovarian CCCs were predominated by the activated dendritic cells, monocytes, CD8 T-cells, and eosinophils. In the uterine CCCs, activated mast cells and regulatory T-cell populations were relatively enriched and they also had unique Wnt/Ca+, oestrogen receptor, and CCR5 signaling (Alldredge et al. 2020).

Serous carcinoma of high grade (HGSOC) and low grade (LGSOC), mucinous carcinomas, clear cell carcinoma and endometrioid carcinoma are the five main subtypes of epithelial ovarian cancer (EOC) ((Prat, D'Angelo, and Espinosa 2018). HGSOC is the most prominent subtype and constitutes about 70% of all the cases. Despite EOC sub-classification, the standard treatments are same for all the subtypes which include cytoreductive surgery and platinum-based chemotherapy with the use of paclitaxel. Deeper interpretation of distinct molecular patterns related to the specific sub-types offers the opportunity to establish personalized cancer treatments. Moreover, understanding of the RNA-Seq data of primary HGSOC tumours and their matched histologically validated HGSOC metastases reveal the differences in the gene expression. Approximately 100 genes showing differential expression were linked with the increased proliferation, decreased embryogenesis & vasculogenesis, and organism's death. This gene signature difference could serve as a marker for cancer prognosis. GATA4, AMHR2 and PCSK6 genes were found to be downregulated in the metastatic samples and were associated with the better prognostic value whereas PAX5 and SFRP2 genes associated with the poorer prognosis were up-regulated (Sallinen et al. 2019). This differential molecular pattern of primary tumours in comparison to metastastatic ones can aid to the improvement in the development of targeted therapies and can nominally predict the patient survival.

Systematic transcriptome analysis performed by Christian L et al. revealed tumour-specific isoforms of ovarian cancer which could be helpful in the early diagnosis and therapy by developing custom bioinformatic algorithms.They analyzed the transcriptome sequence data of 296 high-grade serous ovarian (HGS-OvCa) tumours and 1,839 normal tissues followed by the validation of putative tumour-specific mRNA isoforms by qRT-PCR.Various mRNA candidates with unique sequences or tumour-specific expressions have been identified for the clinical use having role in the diagnosis and therapy. The candidates which were expressed only in the examined cancer tissues but not in the normal tissues include CD9, RAB11FIP4, ETV4, FOXM1, LSR and FGFRL1. This specific expression allows them to be used as therapeutic targets which can be targeted by using monoclonal antibodies and T-cell–based vaccines (Barrett et al. 2015).

Head and neck squamous cell carcinoma (HNSCC) is the cancer having many sub-sites including oral cavity, hypopharynx, larynx and oropharynx, but it is regarded as a single entity due to a common epithelial precursor. Though in reality, it holds complexity and heterogeneity. Development in the web-based tools and informative repository databases provide a clear picture of differently expressed molecules. Along with the mRNA, long non-coding RNAs (lncRNAs) are also included under transcriptomics. Rock et al. analyzed the independent dataset of HNSCC and identified 146 head and neck-specific novel miRNAs and also characterized the tissue-specific transcript expression that helps in deciding the patient's outcome. miR-125, miR-2 and miR-30 are the three identified HNnov-miRNAs which were strongly linked with the HPV status. Downregulation of these miRNAs signifies the presence of HPV infection. Along with this, it has been found that the poor survival is associated with the increased expression of miR-21, miR-18a, miR-155, miR-19a, miR-181a,miR-210 and miR-134a, whereas low expression level of miR-491-p5, miR-451, miR-34a, let-7d, miR-126a, let-7g, miR-17, miR-125b, miR-153, miR-200c, miR-203, miR-363, miR-205, miR 218, and miR-375 were observed and the decreased expression was associated with the poor prognosis (Rock et al. 2019).

miRNAs control approximately 30% of all the human genes, therefore, search of miRNAs and their potential target mRNA involved in the tumour progression could be helpful in the identification of comprehensive miRNA-mRNA regulatory pathway that can be directly exploited for the therapeutic purposes. Integrated molecular biology and bioinformatics analysis of four independent HNSCC datasets were screened out and a couple of miRNA-mRNA regulatory pairs were identified as critical mediators. They also serve as reliable biomarkers for the diagnostic and prognostic purposes of HNSCC. FN1, CXCL5, MMP3, ISGI5, SPP1, LOX, CXCL11, IFIT3, RSAD2 and STAT1 are the ten hub genes and potential biomarkers associated with the HNSCC progression, indicating the oncogenic role of these genes and their role as therapeutic targets (Jin and Qin 2020).

Immune cells in the tumour microenvironment decide the outcome of anti-cancer therapies including immune-therapy. PVR, TNFRSF12A, IL21R,and SOCS1 are immune-related genetic signatures which have been identified from the transcriptomic data analysis of various HNSCC patient cohorts. These signatures can be used to guide the prediction of tumour progression and prognosis in the patients. PVR is a highly expressed gene associated with the poor survival whereas the upregulation of IL21R and SOCS1 is linked with the improved survival. PVR and TNFRSF12A are the genes of poor prognostic value and were negatively correlated with the activated $CD4^{+}$T cells, $CD8^{+}$T cells, follicular helper T cells, and regulatory T cells (Tregs), whereas IL21R and SOCS1 were positively correlated with the above mentioned cell types, partially reflecting the state of the immune microenvironment in HNSCC. The increased levels of these four immune-cell types help in the better prognosis of the HNSCC patients. Twenty two different immune-cell types were estimated through CIBERSORT algorithm and an enhanced proportion of the infiltrating immune cells in the HNSCC tumours was found. There is a correlation between the prognostic factors and the immune-cell types and this correlation confirms the role of these genes in managing the immune response during HNSCC (Yao et al. 2020).

Immunotherapy and nivolumab treatment benefits only a minority of patients with squamous cell carcinoma of the head and neck (SCCHN). It has been found through RNA-seq data that nivolumab treatment alone or in combination with tadalafil induces a differential intratumoral immune response in HPV(+) as well as in HPV(-) SCCHNs. Both HPV(+) and HPV (-) SCCHN patients treated with nivolumab showed increased levels of a wide range of immune-related genes and this effect was amplified by tadalafil. Along with this, enhanced myeloid cell infiltration was found in both the HPV(+) SCCHNs and the HPV(-) SCCHNs, though in HPV(+) SCCHNs, T cell gene expression signatures were more pronounced. These transcript patterns challenge the fact that the increased level of intra-tumoral T lymphocytes alone is linked with the favourable therapeutic outcomes (Shukla et al. 2019).

Immune-related genes, IRGs possessing prognostic value and acting as potential biomarkers have been identified by Li et al. Due to the differential expression of IRGs, normal and cancer cells show distinct range of infiltration of various immune cell subtypes which impacts tumour prognosis and response to the immunotherapy. These one of a kind subsets of tumour-infiltrating cells are correlated with the development and promotion of the tumour. Analysis of these different IRGs could impart deep knowledge of immunology and delineate the subgroups of patients who are more likely to respond to the immunotherapy. Correlation has been reported between IRGs and HNSCC prognosis. Differentially expressed IRGs are linked with patient's prognosis and their biological function is related to the immunocyte migration and receptor-ligand activity. Among the differentially expressed IRGs, the cytokine-cytokine receptor interaction pathway is the most commonly enriched IRG in various tumours. The survival-related IRGs are drastically related to the progression of HNSCC. IRGs are significantly associated with the overall survival (OS). A negative correlation was found between the Bcell and $CD4^+$ T cell infiltration levels and the expression of prognosis-related immune genes regardless of HPV status. Depending upon the expression of IRGs which are associated with the survival, a prognosis prediction model has been developed through the multivariable Cox regression analysis which enables an accurate prediction of the survival (Li et al. 2020). The specific subsets of immune cells can be identified by the scRNA-sequencing in the tumour microenvironment which may help in deciding whether a particular tumour will respond to the immunotherapy or not (Schwartzberg et al. 2017).

The scRNA sequencing technology additionally refines the knowledge of intratumoral heterogeneity in head and neck cancer. Oral cavity squamous cell carcinoma (OCSCC) accounts for one-third of all the head and neck cancer cases and laryngeal squamous cell carcinoma (LSCC) is the second most common head and neck HPV-negative cancer. Retrospective cohort study and genomic analysis from TCGA head and neck cancer cohort reveals that OCSCC comprised primarily of mesenchymal and basal tumours whereas LSCC often includes classical and atypical tumours. This gene expression subtype record can be helpful in predicting the expected nodal metastasis and prognosticate survival. Additionally, it can reveal the potential novel biomarkers of nodal metastasis and survival in the HPV(-) head and neck cancer. OCSCC MS subtype was extensively linked with an increased risk of nodal metastasis and predictive of occult nodal metastasis. In LSCC, the CL subtype was associated with the significantly worse overall survival. Gene expression analysis evaluated a connection between the OCSCC mesenchymal subtype and the lymph node metastasis. The mesenchymal subtype, characterized by epithelial to mesenchymal transition, is associated with the nodal metastasis in OCSCC (Zevallos et al. 2019).

6. Personalized Genome-Wide Cancer Transcriptomics Studies

Uhlen et al. analyzed the transcriptomic data obtained by RNA sequencing from 8000 patients representing 17 different cancer types along with the clinical metadata. The data was retrieved from the TCGA database. By this analysis, they recognized suitable prognostic marker genes specific for a particular type of cancer and created individual metabolic models for each patient by correlating the transcriptomic data with the clinical outcomes (patient survival). They reported considerable inter-individual differences within a particular type of cancer. About 2% key genes were found to be common in all the patients irrespective of the type of cancer. For the cancers with high survival rate such as prostate cancer and testicular cancer, less number of prognostic genes were identified while a large no. of prognostic genes were identified for renal and liver cancer. The authors launched a human pathology atlas based on this study which includes the survival plots of the patient cohorts for all the specific genes in various cancer types, along with the list of prognostic genes for each cancer type. This database is helpful for researchers to correlate the survival data with the expression level of a particular gene in a particular type of cancer. This resource also includes the transcriptomic, proteomic and clinical data of patients which could be very helpful for researchers in the development of personalized medicine for cancer patients (Uhlen et al. 2017).

Demircioglu et al. utilized the RNA sequencing data from Pan-Cancer Analysis of Whole Genomes (PCAWG) and The Cancer Atlas (TCGA) database for cancer patients and GTEx database for normal samples and analyzed the role of alternative promoters in a total of 18468 patient samples. Several genes which showed deregulated promoters or exhibited alternative promoters in cancer belonged to already known cancer-related genes such as ß-catenin, which is a proto-oncogene, BID which is a pro-apoptotic gene, SEPT9, TROY which are cancer biomarkers, MLLT1 which is linked to Wilmstumour and AML, CDK4 and PRKACA which are also associated with the different types of cancers. Apart from cancer-related genes, several novel genes were also found to activate alternative promoters in cancer. They also reported that a number of promoters (73-633 for each tissue) were differentially regulated between cancer and normal tissue. Variation in the activity of promoters among different patients was found to be associated with their survival. For example, EML2, ERBB2 and CDKN2A are some of the examples of genes which can predict patient survival based on the use of alternative promoters. The use of alternative promoters was also observed among the different subtypes of cancer as exemplified by breast and kidney cancer (Demircioğlu et al. 2019).

Hoadley et al. also performed the transcriptome analysis (mRNA as well as miRNA expression) along with the analysis of some other parameters such as aneuploidy, DNA hypermethylation and protein arrays from 10000 patient samples corresponding to 33 types of cancer. The data was retrieved from the TCGA database. They performed the molecular clustering of different cancers and explored different aspects of cancer biology in order to aid in the development of future therapeutics (Hoadley et al. 2018).

7. Role of Transcriptomics in the Precision Oncology

The focus of the precision medicine is to examine the efficacy of a disease prevention and therapeutic strategy by means of tailoring personalized remedies primarily, based on an individual's precise genetic lesions, biomarkers, environment, and lifestyle (Reitz 2016). Based on the identification of gene fusions, molecular markers, alternative promoters, splice variants, novel RNAs and non-coding RNAs by analyzing the transcriptome profiling of cancer patients, the personalized medicines can be designed and tested in the clinical trials. Some researchers have tested the precision medicines for the treatment of patients and some have performed clinical trials based on such studies. For example, fibroblast growth factor receptors (FGFRs) were reported to be novel partners in the gene fusion events by the transcriptome sequencing in certain solid tumors (Wu et al. 2013;Borad et al. 2014). This discovery brought about the clinical trials using tyrosine kinase inhibitors inclusive of ponatinib and BGJ398 for treating the cancer patients having such kind of fusions (national clinical trials NCT02272998 and NCT02160041). A study comprised of 102 cancer patients reported clinically important alterations in 46% of the patients on the basis of transcriptome and exome sequencing. As a result of this study, a novel gene fusion including neurotrophic tyrosine kinase receptor type I (NTRK1) was observed in one infant with spindle cell sarcoma and subsequently, crizotinib was given as a treatment strategy which led to a partial response and stable disease (Mody et al. 2015).

8. Omics Databases involvingTranscriptome Studies

8.1. The Cancer Genome Atlas (TCGA)

Along with the genomic data involving several mutations, this database also provides the transcription profile of 10237 patients belonging to 26 types of cancers by using DNA microarray and RNA sequencing techniques (Tomczak, Czerwińska, and Wiznerowicz 2015). This database also provides the information about the survival time and clinical stage of the cancer resulting in the classification of patients according to the pathological grade of cancer. For example, extracellular matrix genes of breast tumour tissue display variation in the expression according to the cancer stage (Hong et al. 2017).

8.2. The Broad Novartis Cancer Cell Line Encyclopedia (CCLE)

This database also provides the gene expression data from 1037 cell lines covering 54000 genes along with the genomic studies. It also includes 24 pharmacological compounds which were screened depending on the omics profile (Barretina et al. 2012). The combinational studies of the pharmacological screens and transcriptome data have contributed significantly to the development of the personalized medicines. For example, in NRAS mutated cells, the gene expression level of AHR was observed to correlate with the efficacy of MEK inhibitor (Barretina et al. 2012).

8.3. NCI60 Database

The database has a collection of the gene expression data for 54000 genes from 59 cell lines along with the genomics and proteomics data. It also has a collection of 77000 chemicals screened for their anti-cancer effect. This database provides the mutation data, gene expression data and drug response data from the same cell line (Shoemaker 2006). The gene expression studies from this database can be used to categorize cancer into different subtypes and to correlate the gene expression studies with the drug response as exemplified by the discovery of ABCG2 related transporter (Ross et al. 2000; Deeken et al. 2009).

8.4. The Genomics of Drug Sensitivity in Cancer (GDSC)

This database has information about the genomics, transcriptomics and drug screening data from various cancer samples and cell lines. The genomics and transcriptomics data have been correlated with the drug screening data (Iorio et al. 2016).

8.5. The Library of Integrated Network-Based Cellular Signatures (LINCS)

It provides a signature of the transcriptome data for 1000 genes and the screening data for 24000 drugs screened against this transcriptome profile in 53 cell lines. This database also describes the cellular phenotypes such as cell proliferation, etc. along with the protein binding studies (Vempati et al. 2014).

8.6. Project Achilles

This database provides data regarding shRNA (54000 probes) and CRISPR-Cas9 (120000 sgRNA probes) screening in various cell lines to identify the oncogenes (Cowley et al. 2014; Aguirre et al. 2016).

9. Transcriptome Engineering Tools for Cancer Diagnosis and Precision Oncology

9.1. SiRNA/ShRNA

SiRNA and ShRNA have been extensively studied for their role in the diagnosis and treatment of cancer, but they have certain limitations such as limited uptake, off-target effects, immunogenicity, and instability. Different delivery methods have been developed to reduce these limitations and to enhance their utility in the cancer treatment. These screens have identified a number of molecular markers as exemplified by the discovery of various novel members of DNA damage pathway in an osteosarcoma cell line (Rameseder et al. 2015). Another example includes the identification of a correlation between the alterations in ATP1A1 gene with the mutational status of STK11 gene in lung carcinoma (Kim et al. 2016).

Specific targeting of oncogenes by si/shRNA for the treatment of cancer represents a way forward in the path of precision oncology. Several studies have screened these RNAi tools for their anti-cancer effect. For example, in mice xenografted with the keratin forming epidermal carcinoma, siRNA against MCL1 was delivered through the liposome along with the chemotherapeutics and this resulted in the decreased MCL-1 expression (Shim et al. 2011). Neutral nanoliposome (DOPC) mediated delivery of siRNAs targeting FAK, EphA2 and neuropilin-2 genes was found to inhibit the tumor growth in mice xenografted with the colorectal cancer cells and in an orthotopic model of ovarian carcinoma (Landen et al. 2005; Halder et al. 2006; Gray et al. 2008). CALLA-01 is a cyclodextrin-based nanoparticle system containing siRNA against M2 subunit of ribonucleotide reductase. This system showed tumour growth inhibition in the phase I clinical trial studies (Davis et al. 2010). Polyethylenimine (PEI) mediated delivery of siRNA specific to HER2 receptor was found to have an anti-cancer effect in mice (Urban-Klein et al. 2005). Poly(amidoamine) (PAMAM) dendrimer carrying siRNA specific to TWIST1 exhibited decreased invasion and migration in the SUM1315 triple-negative breast cancer cells (TNBC) and xenografted orthotopic tumours (Finlay et al. 2015). McNamara et al. designed a cell-specific siRNA against BCL2 and plk1 by conjugating siRNA to an aptamer which ensures the selective targeting of the prostate cancer cells as it selectively binds to the prostate-specific membrane antigen (PSMA) which is a receptor with high expression in the prostate cancer cells. This siRNA induced cell death in the prostate cancer cells and also exhibited tumour growth inhibition and regression in mice xenografted with the prostate cancer cells (McNamara et al. 2006). A similar approach was used by Geffen et al., an aptamer targeting EpCAM was linked to the siRNA targeting plk1. This chimera inhibited the tumour growth of $EpCAM^+$ breast cancer cells in *invitro* as well as *in vivo* conditions but not in $EpCAM^-$ cells (Gilboa-Geffen et al. 2015).

9.2. CRISPR-Cas 9 System

Clustered regularly interspersed short palindromic repeats (CRISPR)-Cas9 system is being widely utilized in different fields including cancer. Loss of function (LOF) and gain of function (GOF) screens using catalytically dead Cas9 (dCas9) are the potential tools for the identification of cancer biomarkers which can be specifically targeted for the precision oncology. LOF screens involve the fusion of dCas9 to a transcription repressor domain such as KRAB. GOF screens involve the fusion of dCas9 to a single transcription activator domain such as VP64 or multiple activator domains such as VP64-65-Rta. CRISPR screens are more sensitive and effective in identifying the cancer biomarkers as compared to siRNA or shRNA screens. So, large-scale screens including different cancer types have been done using CRISPR-dCas9. For example, 1500 genes were identified in a genome-wide screen among 10 different cancer cell lines (Zhan et al. 2019). Similarly, a study including screening of 8 cell lines of acute myeloid leukemia (AML) resulted in the identification of 429 candidate genes. Out of these genes, KAT2A was recognized as a leukemia-specific targetable biomarker. CRISPR screen in the regulatory region suggested that AML cell lines which have a mutation in the NPM1 gene are dependent on the menin binding position of MLL1 protein indicating the therapeutic potential of the inhibitors of this site (Tzelepis et al. 2016).

Genome-wide CRISPR screen resulted in the identification of MLLT1 (ENL) gene as an essential gene for tumour growth and proliferation in the MLL-AF4 positive AML cell lines (Erb et al. 2017). In melanoma cells, the knockout library generated by CRISPR screens aided in the recognition of the genes required for vemurafenib resistance (Shalem, Sanjana, and Zhang 2015).

Targeting of well-established biomarkers of cancer by using CRISPR–cas9 is an emerging field in designing precision medicines. Several studies have already shown the potential of this strategy in cancer therapeutics. For example, targeting of HER2 by the CRISPR-Cas9 system inhibited the cell growth by deregulating MAPK/ERK signaling (Wang and Sun 2017). CRISPR-Cas9 targeting of E6 and E7 promoters and transcripts substantially inhibited the tumour growth and proliferation in *invitro* as well as *invivo* model of human papilloma virus16-positive cervical cancer cells by increasing the expression of p53 and p21 (Zhen et al. 2014). CRISPR-Cas system can be utilized to knockout the genes which are responsible for glioblastoma such as NFi, Trp53 and Pten (Ratan et al. 2018). Growth and metastasis of TNBC were found to be inhibited by CRISPR-Cas9 mediated knockout of Cripto-1 marker (Ratan et al. 2018). Reversal of the chemo-resistance to cisplatin and paclitaxel was reported by the CRISPR-Cas9 mediated knockdown of HE4 biomarker in the ovarian cancer cells (Ribeiro et al. 2016). shRNA and CRISPR-Cas9 mediated knockdown of mannose receptor, LY75 was found to reduce the invasion and migration in *invitro* as well as *in vivo* experimental set-up of ovarian cancer (Faddaoui et al. 2016).

9.3. CRISPR-Cas13

CRISPR-Cas13 system belongs to class 2 type VI CRISPR-Cas system and can naturally cleave ssRNA as compared to class 2 type II CRISPR-Cas9 system which naturally targets double-strand DNA but can be engineered to cleave ssRNA. This system is a flexible platform for the study of transcriptome. Cox et al. engineered Cas 13b in such a way that it can bind to RNA but lacks endonuclease activity. Further, they fused Cas13b with ADAR2 deaminase which can convert adenosine to inosine to make REPAIR (RNA Editing for Programmable Adenosine to Inosine Replacement) system. By this strategy, they successfully accomplished a precise alteration in a single nucleotide base of RNA, thus paving a way for reverting disease-causing mutations which occur in various diseases including cancer (Cox et al. 2017).

Thus, it could be a powerful tool for the precision oncology as it can be used for cancer diagnosis and treatment without affecting the transcriptome and without any permanent change in the genome. This system can also be used to revert the stop codon leading to the correction of missense mutation and to revert miRNA mutation or polymorphism, both mechanisms are helpful in the anti-cancer therapy.

Based on Cas13a, Gootenberg et al. designed an *invitro* detection system called SHERLOCK (Specific High Sensitivity Enzymatic Reporter Unlocking) which can detect a single base alteration in single strand DNA and RNA (Gootenberg et al. 2017). Further, in another study, Gootenberg et al. attempted to test the usefulness of SHERLOCK as a diagnostic (genotype editing) as well as therapeutic (RNA editing) tool using the REPAIR system to revert the mutations in the HEK293FT cells. They found that 43% editing was successfully achieved and this editing was detectable by multiplexed SHERLOCK (Gootenberg et al. 2018). Catalytically inactive Cas13 conjugated with a fluorescent protein can be utilized to locate the endogenous RNA and their trafficking inside the cell. This approach has a significant role in the diagnosis of cancer as RNA trafficking helps in the determination of the asymmetric distribution of proteins and also regulates the asymmetrical cellular organization. The disruption in this regulatory pathway can lead to tumorigenesis (Huang and Muthuswamy 2010). For example, polarized architecture along the apical and basal axis which is maintained by the cell junctions and specific membrane domains is reported to be dysregulated in a number of epithelial cancers (Huang and Muthuswamy 2010). A deactivated Cas13 system can also be utilized to identify the RNA binding protein (RBP) bound RNA in cancer cell lysates using RNA immunoprecipitation followed by sequencing (Mahas, Neal Stewart, and Mahfouz 2018).

10. Transcriptomics Integrated with Proteomics Show Major Advantages and Identify Novel Therapeutic Vulnerabilities

Though transcriptomic profiling is one of the most utilized approaches for the identification of the root cause of any disease at the molecular level, its integration with the proteomics better resolves the complexity of the disease. By using such approach, scientists can propose more precise diagnostics based on the identified molecular targets. It also helps in finding new metabolic and therapeutic vulnerabilities which was not discovered by the genomic and transcriptomic studies (Krug et al. 2020; Gillette et al. 2020). It can help in the better understanding of the pathways responsible for the therapeutic resistance (Krug et al. 2020). Such studies also aid in the classification of cancers which can reveal new aspects of the cancer therapeutics, for example, proteogenomic sub-typing of cancer can disclose heterogeneity among tumours of the same stage which in-turn help in the identification of aggressive clinical features (Chen et al. 2020). Subtype specificity of the phosphorylated kinases was also reported in breast cancer by a proteogenomic based classification resulting in an improved targetability (Krug et al. 2020).

Profiling of tumours by integrating the DNA and RNA sequencing data with mass-spectroscopy-based proteomic data allows more comprehensive analysis of tumours. Krug et al. performed this type of data integration. Proteogenomic analysis of 122 naive primary breast cancer tissues led to a clear view of the breast cancer biology including a detailed characterization of ERBB2 amplicon, immunogenicity of the tumour microenvironment, heterogeneity in the expression of targetable markers, multi-omics categorization and metabolic profiling of breast cancer. In addition to this, the study explored the proteogenomic status of Rb gene with more accuracy which has clinical importance in predicting the responsiveness of cdk4/6 inhibitors against breast cancer. Post-translational modifications such as phosphorylation and acetylation were also analyzed resulting in the determination of the expression status of phosphorylated kinases in a subtype-specific manner and acetylation status of DNA damage response proteins, respectively (Krug et al. 2020).

Gillette et al. identified various therapeutic vulnerabilities based on a proteogenomic study in lung adenocarcinoma which include anti-CTLA4 therapy and inhibition of IDO1 in immune hot tumors. They also found neutrophyll degranulation as a potential immune-modulatory mechanism in lung adenocarcinoma having the STK1 mutation. This observation was only evident in the proteomic study and was not detected at mRNA level indicating the importance of integrating proteomics with the genomic and transcriptomic studies. Some other therapeutical vulnerabilities reported in this study included the inhibition of SOS in KRAS mutant and inhibition of PTPN11(Shp2) in ALK fusion as well as EGFR mutant tumors of which PTPN11 is a well-established therapeutic target and is undergoing the clinical trials (Gillette et al. 2020).

Mertins et al. discovered a subgroup of G-protein-coupled receptor in breast cancer through proteomic analysis which was not detectable at mRNA level. Other therapeutic vulnerabilities for breast cancer were also revealed in this proteogenomic study including certain targetable kinases such as CDK12, TLK2, PAK1 and RIPK2 (Mertins et al. 2016). In a study comprising of clear cell renal cell carcinoma, a protein signature was revealed by proteogenomic analysis. The proteins which were upregulated included proteins indulged in the epithelial to mesenchymal transition (EMT), immune response proteins and proteins involved in different signaling pathways, for instance, glycolysis, mTOR and hypoxia, etc. The proteins which were downregulated belonged to fatty acid metabolism, oxidative phosphorylation, and tricarboxylic acid (TCA) cycle. Downregulation of TCA cycle and oxidative phosphorylation proteins were only evident by proteomic analysis and not detected at mRNA level (Clark et al. 2019).

Proteogenomic profiling of head and neck cancer (Huang et al. 2021) and pediatric brain cancer (Petralia et al. 2020) added an extra step in the direction of precision medicine development by improving the sub-classification of these cancers based on the proteomic profiling which aided in the precise prediction of the therapeutic response of a particular inhibitor for different cancer subtypes. Proteogenomic landscape of lung cancer in East Asia revealed demographic differences in the driver mutation frequencies of therapeutically important genes between Taiwan (TW) and the cancer genome atlas (TCGA) lung adenocarcinoma cohorts as exemplified clearly by RBM10 gene. Furthermore, authors have reported high inter-patient differences in the phosphorylation sites on Raf/MEK/ERK pathway. High phosphorylation status of the proteins of MAPK pathway was found to be associated with the mutations in the EGFR and KRAS gene while low phosphorylation was reported to be associated with the mutations in the TP53 gene. Thus, proteomic profiling explored the heterogeneity within the tumors of same stage. Additionally, APOBEC mutations were observed to be the driving factor for early onset of lung adenocarcinoma in females and their high protein and phosphorylation levels make them good candidates for therapeutic targeting (Chen et al. 2020). A proteogenomic study by Stewart et al. reported certain metabolic vulnerabilities in one of the subtypes of squamous cell lung carcinoma. They sub-classified squamous cell lung carcinoma into three subtypes based on the proteomic data. These subtypes are inflamed, redox and mixed. TP63, TFRC and PSAT1 were identified as metabolic vulnerability in the redox subtype (Stewart et al. 2019).

11. Limitations of the Single Omic (Transcriptome) Based Study

This kind of study represents a good resource for investigation of cancer and in the development of precision medicine, but such studies have various inherent limitations. Different factors which affect cancer development such as geography, ethnicity, gender, mutational factors and other lifestyle factors such as smoking show interdependence on each other and thus, the interpretation based on any of these single factors needs to be considered with caution. Linear modeling for adjusting this interdependence may not be successful owing to the large data set. For example, an association analysis for gender and smoking effects was reported to be erroneous (Gillette et al. 2020). This kind of effort may lead to extraneous information about the cancer biology due to less resolution. Additionally, the role of tumour microenvironment is not very clear in such cases. Thus, integrating the single cell genomics and transcriptomics with proteomics would provide more spatial resolution and might give detailed information about the role of tumour microenvironment in the tumour development.

Transcriptomics has contributed hugely in the field of cancer diagnostics and therapeutics, but there are several other complexities associated with cancer biology which need to be considered. Novel omic approaches such as epigenomics, metabolomics, proteomics and lipidomics have added an extra layer of information about the cancer biology. Epigenomics research has revealed the role of methylation, acetylation, phosphorylation and other modifications of DNA and histone proteins in cancer development which cannot be explained by the genomic and transcriptomic based studies. Although genomic and transcriptomic alterations make the base for proteomics, but proteomics data is more practical as it has been found that correlation between mRNA abundance and protein abundance can be poor (Stewart et al. 2019). Lipidomics and metabolomics also increase the tumour complexity indicating towards the importance of integration of different omic approaches rather than relying on just one approach.

Moreover, drug resistance mechanisms, tumour heterogeneity, stem cell properties of cancer cells and the tumour microenvironment are few more challenges that need to be addressed while translating the molecular characterization of cancer into the development of precision medicines for its treatment. Thus, the integration of multiple omic approaches would be the best choice for the precision oncology.

Conclusion and Future Prospects

Transcriptomics overcomes the hurdles related to the genomic studies such as heterogeneity and complexity, hence representing an effective, simple and robust tool for target-biomarker identification for designing precision medicine. Transcriptomics plays a multifaceted role in the field of precision oncology. Gene expression studies revealed different aspects of cancer biology, aided in the identification of diagnostic, prognostic and predictive cancer biomarkers, cancer-related gene fusions, classification of cancers and identification of the tissue of origin of cancers with unknown primary tumour origin. RNA sequencing studies also contributed to uncover the role of alternative promoters in the cancer progression and patient survival. It also contributed to the prediction of clinical outcomes (cancer prognosis), designing of targeted/personalized therapeutics by means of designing inhibitors against specific cancer biomarkers or through gene silencing therapies. Thus, transcriptome based molecular signature and editing tools such as the CRISPR-Cas system, si/shRNA tools play a significant role in the precision oncology.

Despite of the huge potential of transcriptome profiling in the field of precision oncology, there are various challenges in the development of the precision medicine for cancer treatment. Other approaches in omics add to the complexity of cancer biology such as epigenomics also regulates the gene expression irrespective of genomic and transcriptomic alterations. Proteomics and metabolomics represent other aspects of cancer development. Thus, for the development of an effective personalized cancer treatment, multiomic approaches have to be considered. We envision the integration of transcriptomics with genomics, proteomics, epigenomics and metabolomics to get deep insight into the tumour biology. This could help in the identification of therapeutically important molecular targets for the precision oncology along with the development of improved diagnostic and prognostic markers. Furthermore, combination therapies targeting multiple molecular markers identified through transcriptomics and other omic approaches should be encouraged instead of monotherapy. Other therapeutic approaches such as immunotherapy, use of oncolytic viruses, inhibitors targeting cancer epigenomics, metabolomics, lipidomics and stem cells along with the common therapeutic inhibitors should be utilized with the multidisciplinary collaboration among the specialists of different fields.

References

Adamia, Sophia, Benjamin Haibe-Kains, Patrick M Pilarski, Michal Bar-Natan, Samuel Pevzner, Herve Avet-Loiseau, Laurence Lode, Sigitas Verselis, Edward A Fox, and John Burke. 2014. "A Genome-Wide Aberrant RNA Splicing in Patients with Acute Myeloid Leukemia Identifies Novel Potential Disease Markers and Therapeutic Targets." *Clinical Cancer Research*, 20 (5), 1135–45. https://doi:10.1158/1078-0432.CCR-13-0956.

Aguirre, Andrew J., Robin M. Meyers, Barbara A. Weir, Francisca Vazquez, Cheng-Zhong Zhang, Uri Ben-David, April Cook, et al. 2016. "Genomic Copy Number Dictates a Gene-Independent Cell Response to CRISPR/Cas9 Targeting." *Cancer Discovery*, 6 (8), 914–29. https://doi.org/10.1158/2159-8290.CD-16-0154.

Akino, Tomoshige, Kyoko Hida, Yasuhiro Hida, Kunihiko Tsuchiya, Deborah Freedman, Chikara Muraki, Noritaka Ohga, et al. 2009. "Cytogenetic Abnormalities of Tumor-Associated Endothelial Cells in Human Malignant Tumors." *The American Journal of Pathology*, 175 (6), 2657–67. https://doi.org/10.2353/ajpath.2009.090202.

Alizadeh, A. A., Eisen, M. B., Davis, R. E., Ma, C., Lossos, I. S., Rosenwald, A., Boldrick, J. C., et al. 2000. "Distinct Types of Diffuse Large B-Cell Lymphoma Identified by Gene Expression Profiling." *Nature*, 403 (6769), 503–11. https://doi.org/10.1038/35000501.

Alldredge, Jill, Leslie Randall, Gabriela De Robles, Anshu Agrawal, Dan Mercola, Marisa Liu, Pavneet Randhawa, Robert Edwards, Michael McClelland, and Farah Rahmatpanah. 2020. "Transcriptome Analysis of Ovarian and Uterine Clear Cell Malignancies." *Frontiers in Oncology*, 10. https://doi.org/10.3389/fonc.2020.598579.

Barretina, Jordi, Giordano Caponigro, Nicolas Stransky, Kavitha Venkatesan, Adam A. Margolin, Sungjoon Kim, Christopher J. Wilson, et al. 2012. "The Cancer Cell Line Encyclopedia Enables Predictive Modelling of Anticancer Drug Sensitivity." *Nature*, 483 (7391), 603–7. https://doi.org/10.1038/nature11003.

Barrett, Christian L., Christopher DeBoever, Kristen Jepsen, Cheryl C. Saenz, Dennis A. Carson, and Kelly A. Frazer. 2015. "Systematic Transcriptome Analysis Reveals Tumor-Specific Isoforms for Ovarian Cancer Diagnosis and Therapy." *Proceedings of the National Academy of Sciences*, 112 (23), E3050–57. https://doi.org/10.1073/pnas.1508057112.

Barton, Sarah, Eliza A. Hawkes, Andrew Wotherspoon, and David Cunningham. 2012. "Are We Ready to Stratify Treatment for Diffuse Large B-Cell Lymphoma Using Molecular Hallmarks?" *The Oncologist*, 17 (12), 1562–73. https://doi.org/10.1634/theoncologist.2012-0218.

Bartoschek, Michael, NikolayOskolkov, Matteo Bocci, John Lövrot, Christer Larsson, Mikael Sommarin, Chris D. Madsen, et al. 2018. "Spatially and Functionally Distinct Subclasses of Breast Cancer-Associated Fibroblasts Revealed by Single Cell RNA Sequencing." *Nature Communications*, 9 (December), https://doi.org/10.1038/s41467-018-07582-3.

Borad, Mitesh J., Mia D. Champion, Jan B. Egan, Winnie S. Liang, Rafael Fonseca, Alan H. Bryce, Ann E. McCullough, et al. 2014. "Integrated Genomic Characterization Reveals Novel, Therapeutically Relevant Drug Targets in FGFR and EGFR Pathways in Sporadic Intrahepatic Cholangiocarcinoma." *PLoS Genetics*, 10 (2), e1004135. https://doi.org/10.1371/journal.pgen.1004135.

Casamassimi, Amelia, Antonio Federico, Monica Rienzo, Sabrina Esposito, and Alfredo Ciccodicola. 2017. "Transcriptome Profiling in Human Diseases: New Advances and Perspectives." *International Journal of Molecular Sciences*, 18 (8), 1-15. https://doi.org/10.3390/ijms18081652.

Chen, Yi-Ju, Theodoros I Roumeliotis, Ya-Hsuan Chang, Ching-Tai Chen, Chia-Li Han, Miao-Hsia Lin, Huei-Wen Chen, Gee-Chen Chang, Yih-Leong Chang, and Chen-Tu Wu. 2020. "Proteogenomics of Non-Smoking Lung Cancer in East Asia Delineates Molecular Signatures of Pathogenesis and Progression." *Cell*, 182 (1), 226–44.

Clark, David J, Saravana M Dhanasekaran, Francesca Petralia, Jianbo Pan, Xiaoyu Song, Yingwei Hu, Felipe da Veiga Leprevost, Boris Reva, Tung-Shing M Lih, and Hui-Yin Chang. 2019. "Integrated Proteogenomic Characterization of Clear Cell Renal Cell Carcinoma." *Cell*, 179 (4), 964–83.

Cowley, Glenn S., Barbara A. Weir, Francisca Vazquez, Pablo Tamayo, Justine A. Scott, Scott Rusin, Alexandra East-Seletsky, et al. 2014. "Parallel Genome-Scale Loss of Function Screens in 216 Cancer Cell Lines for the Identification of Context-Specific Genetic Dependencies." *Scientific Data*, 1, 140035. https://doi.org/10.1038/sdata.2014.35.

Cox, David B. T., Jonathan S. Gootenberg, Omar O. Abudayyeh, Brian Franklin, Max J. Kellner, Julia Joung, and Feng Zhang. 2017. "RNA Editing with CRISPR-Cas13." *Science (New York, N.Y.)*, 358 (6366), 1019–27. https://doi.org/10.1126/science.aaq0180.

Davis, Mark E., Jonathan E. Zuckerman, Chung Hang J. Choi, David Seligson, Anthony Tolcher, Christopher A. Alabi, Yun Yen, Jeremy D. Heidel, and Antoni Ribas. 2010. "Evidence of RNAi in Humans from Systemically Administered SiRNA via Targeted Nanoparticles." *Nature*, 464 (7291), 1067–70. https://doi.org/10.1038/nature08956.

Deeken, John F., Robert W. Robey, Suneet Shukla, Kenneth Steadman, Arup R. Chakraborty, Balasubramanian Poonkuzhali, Erin G. Schuetz, Susan Holbeck, Suresh V. Ambudkar, and Susan E. Bates. 2009. "Identification of Compounds That Correlate with ABCG2 Transporter Function in the National Cancer Institute Anticancer Drug Screen." *Molecular Pharmacology*, 76 (5), 946–56. https://doi.org/10.1124/mol.109.056192.

Demircioğlu, Deniz, EnginCukuroglu, Martin Kindermans, Tannistha Nandi, Claudia Calabrese, Nuno A. Fonseca, André Kahles, et al. 2019. "A Pan-Cancer Transcriptome Analysis Reveals Pervasive Regulation through Alternative Promoters." *Cell*, 178 (6), 1465-1477.e17. https://doi.org/10.1016/j.cell.2019.08.018.

Ellis, Matthew J., and Charles M. Perou. 2013. "The Genomic Landscape of Breast Cancer as a Therapeutic Roadmap." *Cancer Discovery*, 3 (1), 27–34. https://doi.org/10.1158/2159-8290.CD-12-0462.

Erb, Michael A., Thomas G. Scott, Bin E. Li, Huafeng Xie, Joshiawa Paulk, Hyuk-Soo Seo, Amanda Souza, et al. 2017. "Transcription Control by the ENL YEATS Domain in Acute Leukaemia." *Nature*, 543 (7644), 270–74. https://doi.org/10.1038/nature21688.

Faddaoui, Adnen, Magdalena Bachvarova, Marie Plante, Jean Gregoire, Marie-Claude Renaud, Alexandra Sebastianelli, Stephane Gobeil, et al. 2016. "The Mannose Receptor LY75 (DEC205/CD205) Modulates Cellular Phenotype and Metastatic Potential of Ovarian Cancer Cells." *Oncotarget*, 7 (12), 14125–42. https://doi.org/10.18632/oncotarget.7288.

Finlay, James, Cai M. Roberts, Gina Lowe, Joana Loeza, John J. Rossi, and Carlotta A. Glackin. 2015. "RNA-Based TWIST1 Inhibition via Dendrimer Complex to Reduce Breast Cancer Cell Metastasis." *BioMed Research International*, 2015, 382745. https://doi.org/10.1155/2015/382745.

Gilboa-Geffen, Adi, Peter Hamar, Minh T. N. Le, Lee Adam Wheeler, Radiana Trifonova, Fabio Petrocca, Anders Wittrup, and Judy Lieberman. 2015. "Gene Knockdown by EpCAM Aptamer-SiRNA Chimeras Suppresses Epithelial Breast Cancers and Their Tumor-Initiating Cells." *Molecular Cancer Therapeutics*, 14 (10), 2279–91. https://doi.org/10.1158/1535-7163.MCT-15-0201-T.

Gillette, M. A., Satpathy, S., Cao, S., Dhanasekaran, S. M., Vasaikar, S. V., Krug, K., Petralia, F., Li, Y., Liang, W. W. and Reva, B. 2020. "Clinical Proteomic Tumor Analysis Consortium Proteogenomic Characterization Reveals Therapeutic Vulnerabilities in Lung Adenocarcinoma." *Cell*, 182 (1), 200–225.

Gootenberg, Jonathan S., Omar O. Abudayyeh, Jeong Wook Lee, Patrick Essletzbichler, Aaron J. Dy, Julia Joung, Vanessa Verdine, et al. 2017. "Nucleic Acid Detection with CRISPR-Cas13a/C2c2." *Science (New York, N.Y.)*, 356 (6336), 438–42. https://doi.org/10.1126/science.aam9321.

Gootenberg, Jonathan S., Omar O. Abudayyeh, Max J. Kellner, Julia Joung, James J. Collins, and Feng Zhang. 2018. "Multiplexed and Portable Nucleic Acid Detection Platform with Cas13, Cas12a, and Csm6." *Science (New York, N.Y.)*, 360 (6387), 439–44. https://doi.org/10.1126/science.aaq0179.

Goveia, Jermaine, Katerina Rohlenova, Federico Taverna, Lucas Treps, Lena-Christin Conradi, Andreas Pircher, Vincent Geldhof, et al. 2020. "An Integrated Gene Expression Landscape Profiling Approach to Identify Lung Tumor Endothelial Cell Heterogeneity and Angiogenic Candidates." *Cancer Cell*, 37 (1), 21-36.e13. https://doi.org/10.1016/j.ccell.2019.12.001.

Granados-Riveron, Javier T., and Guillermo Aquino-Jarquin. 2018. "CRISPR-Cas13 Precision Transcriptome Engineering in Cancer." *Cancer Research*, 78 (15), 4107–13. https://doi.org/10.1158/0008-5472.CAN-18-0785.

Gray, Michael J., George Van Buren, Nikolaos A. Dallas, Ling Xia, Xuemei Wang, Anthony D. Yang, Ray J. Somcio, et al. 2008. "Therapeutic Targeting of Neuropilin-2 on Colorectal Carcinoma Cells Implanted in the Murine Liver." *Journal of the National Cancer Institute*, 100 (2), 109–20. https://doi.org/10.1093/jnci/djm279.

Hainsworth, John D., Mark S. Rubin, David R. Spigel, Ralph V. Boccia, Samuel Raby, Raven Quinn, and F. Anthony Greco. 2013. "Molecular Gene Expression Profiling to Predict the Tissue of Origin and Direct Site-Specific Therapy in Patients with Carcinoma of Unknown Primary Site: A Prospective Trial of the Sarah Cannon Research Institute." *Journal of Clinical Oncology: Official Journal of the American Society of Clinical Oncology*, 31 (2), 217–23. https://doi.org/10.1200/JCO.2012.43.3755.

Halder, Jyotsnabaran, Aparna A. Kamat, Charles N. Landen, Liz Y. Han, Susan K. Lutgendorf, Yvonne G. Lin, William M. Merritt, et al. 2006. "Focal Adhesion Kinase Targeting Using *In Vivo* Short Interfering RNA Delivery in Neutral Liposomes for Ovarian Carcinoma Therapy." *Clinical Cancer Research : An Official Journal of the American Association for Cancer Research*, 12 (16), 4916–24. https://doi.org/10.1158/1078-0432.CCR-06-0021.

Hashimshony, Tamar, NaftalieSenderovich, Gal Avital, Agnes Klochendler, Yaron de Leeuw, Leon Anavy, Dave Gennert, et al. 2016. "CEL-Seq2: Sensitive Highly-Multiplexed Single-Cell RNA-Seq." *Genome Biology*, 17 (April), 77. https://doi.org/10.1186/s13059-016-0938-8.

Hoadley, Katherine A., Christina Yau, Toshinori Hinoue, Denise M. Wolf, Alexander J. Lazar, Esther Drill, Ronglai Shen, et al. 2018. "Cell-of-Origin Patterns Dominate the Molecular Classification of 10,000 Tumors from 33 Types of Cancer." *Cell*, 173 (2), 291-304.e6. https://doi.org/10.1016/j.cell.2018.03.022.

Hong, Yourae, Nayoung Kim, Chao Li, Euna Jeong, and Sukjoon Yoon. 2017. "Patient Sample-Oriented Analysis of Gene Expression Highlights Extracellular Signatures in Breast Cancer Progression." *Biochemical and Biophysical Research Communications*, 487 (2), 307–12. https://doi.org/10.1016/j.bbrc.2017.04.055.

Huang, Chen, Lijun Chen, Sara R Savage, Rodrigo Vargas Eguez, Yongchao Dou, Yize Li, Felipe da Veiga Leprevost, Eric J Jaehnig, Jonathan T Lei, and Bo Wen. 2021. "Proteogenomic Insights into the Biology and Treatment of HPV-Negative Head and Neck Squamous Cell Carcinoma." *Cancer Cell*, 39 (3), 361–79.

Huang, Ling, and Senthil K. Muthuswamy. 2010. "Polarity Protein Alterations in Carcinoma: A Focus on Emerging Roles for Polarity Regulators." *Current Opinion in Genetics & Development*, 20 (1), 41–50. https://doi.org/10.1016/j.gde.2009.12.001.

Iorio, Francesco, Theo A. Knijnenburg, Daniel J. Vis, Graham R. Bignell, Michael P. Menden, Michael Schubert, Nanne Aben, et al. 2016. "A Landscape of Pharmacogenomic Interactions in Cancer." *Cell*, 166 (3), 740–54. https://doi.org/10.1016/j.cell.2016.06.017.

Jaitin, Diego Adhemar, Ephraim Kenigsberg, Hadas Keren-Shaul, Naama Elefant, Franziska Paul, Irina Zaretsky, Alexander Mildner, et al. 2014. "Massively Parallel Single-Cell RNA-Seq for Marker-Free Decomposition of Tissues into Cell Types." *Science (New York, N.Y.)*, 343 (6172), 776–79. https://doi.org/10.1126/science.1247651.

Jin, Yu, and Xing Qin. 2020. "Comprehensive Analysis of Transcriptome Data for Identifying Biomarkers and Therapeutic Targets in Head and Neck Squamous Cell Carcinoma." *Annals of Translational Medicine*, 8 (6), 282–282. https://doi.org/10.21037/atm.2020.03.30.

Kim, Nayoung, Hwa Young Yim, Ningning He, Cheol-Jung Lee, Ju Hyun Kim, Jin-Sung Choi, Hye Suk Lee, et al. 2016. "Cardiac Glycosides Display Selective Efficacy for STK11 Mutant Lung Cancer." *Scientific Reports*, 6 (July), 29721. https://doi.org/10.1038/srep29721.

Klijn, Christiaan, Steffen Durinck, Eric W. Stawiski, Peter M. Haverty, Zhaoshi Jiang, Hanbin Liu, Jeremiah Degenhardt, et al. 2015. "A Comprehensive Transcriptional Portrait of Human Cancer Cell Lines." *Nature Biotechnology*, 33 (3), 306–12. https://doi.org/10.1038/nbt.308

Krug, Karsten, Eric J Jaehnig, Shankha Satpathy, Lili Blumenberg, Alla Karpova, Meenakshi Anurag, George Miles, Philipp Mertins, Yifat Geffen, and Lauren C Tang. 2020. "Proteogenomic Landscape of Breast Cancer Tumorigenesis and Targeted Therapy." *Cell*, 183 (5), 1436–56.

Lambrechts, Diether, ElsWauters, Bram Boeckx, Sara Aibar, David Nittner, Oliver Burton, Ayse Bassez, et al. 2018. "Phenotype Molding of Stromal Cells in the Lung Tumor Microenvironment." *Nature Medicine*, 24 (8), 1277–89. https://doi.org/10.1038/s41591-018-0096-5.

Landen, Charles N., Arturo Chavez-Reyes, Corazon Bucana, Rosemarie Schmandt, Michael T. Deavers, Gabriel Lopez-Berestein, and Anil K. Sood. 2005. "Therapeutic EphA2 Gene Targeting *in Vivo* Using Neutral Liposomal Small Interfering RNA Delivery." *Cancer Research*, 65 (15), 6910–18. https://doi.org/10.1158/0008-5472.CAN-05-0530.

Li, Lei, Xiao-Li Wang, Qian Lei, Chuan-Zheng Sun, Yan Xi, Ran Chen, and Yong-Wen He. 2020. "Comprehensive Immunogenomic Landscape Analysis of Prognosis-Related Genes in Head and Neck Cancer." *Scientific Reports*, 10 (1), 6395. https://doi.org/10.1038/s41598-020-63148-8.

Luo, Wei, Qiang Hu, Dan Wang, Kristin K. Deeb, Yingyu Ma, Carl D. Morrison, Song Liu, Candace S. Johnson, and Donald L. Trump. 2013. "Isolation and Genome-Wide Expression and Methylation Characterization of CD31+ Cells from Normal and Malignant Human Prostate Tissue." *Oncotarget*, 4 (9), 1472–83. https://doi.org/10.18632/oncotarget.1269.

Mahas, Ahmed, C. Neal Stewart, and Magdy M. Mahfouz. 2018. "Harnessing CRISPR/Cas Systems for Programmable Transcriptional and Post-Transcriptional Regulation." *Biotechnology Advances*, 36 (1), 295–310. https://doi.org/10.1016/j.biotechadv.2017.11.008.

Maher, Christopher A., Chandan Kumar-Sinha, Xuhong Cao, Shanker Kalyana-Sundaram, Bo Han, Xiaojun Jing, Lee Sam, Terrence Barrette, Nallasivam Palanisamy, and Arul M. Chinnaiyan. 2009. "Transcriptome Sequencing to Detect Gene Fusions in Cancer." *Nature*, 458 (7234), 97–101. https://doi.org/10.1038/nature07638.

Maher, Christopher A., Nallasivam Palanisamy, John C. Brenner, Xuhong Cao, Shanker Kalyana-Sundaram, Shujun Luo, Irina Khrebtukova, et al. 2009. "Chimeric Transcript Discovery by Paired-End Transcriptome Sequencing." *Proceedings of the National Academy of Sciences of the United States of America*, 106 (30), 12353–58. https://doi.org/10.1073/pnas.0904720106.

McNamara, James O., Eran R. Andrechek, Yong Wang, Kristi D. Viles, Rachel E. Rempel, Eli Gilboa, Bruce A. Sullenger, and Paloma H. Giangrande. 2006. "Cell Type-Specific Delivery of SiRNAs with Aptamer-SiRNA Chimeras." *Nature Biotechnology*, 24 (8), 1005–15. https://doi.org/10.1038/nbt1223.

Meacham, Corbin E., and Sean J. Morrison. 2013. "Tumour Heterogeneity and Cancer Cell Plasticity." *Nature*, 501 (7467), 328–37. https://doi.org/10.1038/nature12624.

Mertins, Philipp, DR Mani, Kelly V Ruggles, Michael A Gillette, Karl R Clauser, Pei Wang, Xianlong Wang, Jana W Qiao, Song Cao, and Francesca Petralia. 2016. "Proteogenomics Connects Somatic Mutations to Signalling in Breast Cancer." *Nature*, 534 (7605), 55–62.

Mody, Rajen J., Yi-Mi Wu, Robert J. Lonigro, Xuhong Cao, Sameek Roychowdhury, Pankaj Vats, Kevin M. Frank, et al. 2015. "Integrative Clinical Sequencing in the Management of Refractory or Relapsed Cancer in Youth." *JAMA*, 314 (9), 913–25. https://doi.org/10.1001/jama.2015.10080.

MRNA-Seq from Single-Cell Levels of RNA and Individual Circulating Tumor Cells. *Nat B*Ziegenhain, Christoph, Beate Vieth, Swati Parekh, Björn Reinius, Amy Guillaumet-Adkins, Martha Smets, Heinrich Leonhardt, Holger Heyn, Ines Hellmann, and Wolfgang Enard. 2017. "Comparative Analysis of Single-Cell RNA Sequencing Methods." *Molecular Cell*, 65 (4), 631-643.e4. https://doi.org/10.1016/j.molcel.2017.01.023.

Muthukkaruppan, V. R., Kubai, L. and Auerbach, R. 1982. "Tumor-Induced Neovascularization in the Mouse Eye." *Journal of the National Cancer Institute*, 69 (3), 699–708.

Nagl, Laurenz, Lena Horvath, Andreas Pircher, and Dominik Wolf. 2020. "Tumor Endothelial Cells (TECs) as Potential Immune Directors of the Tumor Microenvironment – New Findings and Future Perspectives." *Frontiers in Cell and Developmental Biology* 8. https://doi.org/10.3389/fcell.2020.00766.

Paik, Soonmyung, Steven Shak, Gong Tang, Chungyeul Kim, Joffre Baker, Maureen Cronin, Frederick L. Baehner, et al. 2004. "A Multigene Assay to Predict Recurrence of Tamoxifen-Treated, Node-Negative Breast Cancer." *The New England Journal of Medicine*, 351 (27), 2817–26. https://doi.org/10.1056/NEJMoa041588.

Perou, C. M., Sørlie, T., Eisen, M. B., van de Rijn, M., Jeffrey, S. S., Rees, C. A., Pollack, J. R. et al. 2000. "Molecular Portraits of Human Breast Tumours." *Nature*, 406 (6797), 747–52. https://doi.org/10.1038/35021093.

Petralia, Francesca, Nicole Tignor, Boris Reva, Mateusz Koptyra, Shrabanti Chowdhury, Dmitry Rykunov, Azra Krek, Weiping Ma, Yuankun Zhu, and Jiayi Ji. 2020. "Integrated Proteogenomic Characterization across Major Histological Types of Pediatric Brain Cancer." *Cell*, 183 (7), 1962–85.

Prat, Jaime, EmanuelaD'Angelo, and Iñigo Espinosa. 2018. "Ovarian Carcinomas: At Least Five Different Diseases with Distinct Histological Features and Molecular Genetics." *Human Pathology*, 80 (October), 11–27. https://doi.org/10.1016/j.humpath.2018.06.018.

Rameseder, Jonathan, Konstantin Krismer, Yogesh Dayma, Tobias Ehrenberger, Mun Kyung Hwang, Edoardo M. Airoldi, Scott R. Floyd, and Michael B. Yaffe. 2015. "A Multivariate Computational Method to Analyze High-Content RNAi Screening Data." *Journal of Biomolecular Screening*, 20 (8), 985–97. https://doi.org/10.1177/1087057115583037.

Ratan, Zubair Ahmed, Young-Jin Son, Mohammad Faisal Haidere, Bhuiyan Mohammad Mahtab Uddin, Md Abdullah Yusuf, Sojib Bin Zaman, Jong-Hoon Kim, Laila Anjuman Banu, and Jae Youl Cho. 2018. "CRISPR-Cas9: A Promising Genetic Engineering Approach in Cancer Research." *Therapeutic Advances in Medical Oncology*, 10, 1758834018755089. https://doi.org/10.1177/1758834018755089.

Reitz, Christiane. 2016. "Toward Precision Medicine in Alzheimer's Disease." *Annals of Translational Medicine*, 4 (6), 107. https://doi.org/10.21037/atm.2016.03.05.

Ribeiro, J. R., Schorl, C., Yano, N., Romano, N.,Kim, K. K., Singh, R. K. and Moore, R. G. 2016. "HE4 Promotes Collateral Resistance to Cisplatin and Paclitaxel in Ovarian Cancer Cells." *Journal of Ovarian Research*, 9 (1), 28. https://doi.org/10.1186/s13048-016-0240-0.

Roberts, K.G., Li, Y., Payne-Turner, D., Harvey, R. C., Yang, Y.-L., Pei, D.,McCastlain, K. et al. 2014. "Targetable Kinase-Activating Lesions in Ph-like Acute Lymphoblastic Leukemia." *The New England Journal of Medicine*, 371 (11), 1005–15. https://doi.org/10.1056/NEJMoa1403088.

Roberts, Kathryn G., Ryan D. Morin, Jinghui Zhang, Martin Hirst, Yongjun Zhao, Xiaoping Su, Shann-Ching Chen, et al. 2012. "Genetic Alterations Activating Kinase and Cytokine Receptor Signaling in High-Risk Acute Lymphoblastic Leukemia." *Cancer Cell*, 22 (2), 153–66. https://doi.org/10.1016/j.ccr.2012.06.005.

Rock, Leigha D., Brenda C. Minatel, Erin A. Marshall, Florian Guisier, Adam P. Sage, Mateus Camargo Barros-Filho, Greg L. Stewart, Cathie Garnis, and Wan L. Lam. 2019. "Expanding the Transcriptome of Head and Neck Squamous Cell Carcinoma Through Novel MicroRNA Discovery." *Frontiers in Oncology*, 9, https://doi.org/10.3389/fonc.2019.01305.

Rodon, Jordi, Jean-Charles Soria, Raanan Berger, Wilson H. Miller, Eitan Rubin, Aleksandra Kugel, Apostolia Tsimberidou, et al. 2019. "Genomic and Transcriptomic Profiling Expands Precision Cancer Medicine: The WINTHER Trial." *Nature Medicine*, 25 (5), 751–58. https://doi.org/10.1038/s41591-019-0424-4.

Rohlenova, Katerina, Jermaine Goveia, Melissa García-Caballero, Abhishek Subramanian, Joanna Kalucka, Lucas Treps, Kim D. Falkenberg, et al. 2020. "Single-Cell RNA Sequencing Maps Endothelial Metabolic Plasticity in Pathological Angiogenesis." *Cell Metabolism*, 31 (4), 862-877.e14. https://doi.org/10.1016/j.cmet.2020.03.009.

Rosenberg, Alexander B., Charles M. Roco, Richard A. Muscat, Anna Kuchina, Paul Sample, Zizhen Yao, Lucas T. Graybuck, et al. 2018. "Single-Cell Profiling of the Developing Mouse Brain and Spinal Cord with Split-Pool Barcoding." *Science (New York, N.Y.)*, 360 (6385), 176–82. https://doi.org/10.1126/science.aam8999.

Ross, Douglas T, Uwe Scherf, Michael B Eisen, Charles M Perou, Christian Rees, Paul Spellman, Vishwanath Iyer, Stefanie S Jeffrey, Matt Van de Rijn, and Mark Waltham. 2000. "Systematic Variation in Gene Expression Patterns in Human Cancer Cell Lines." *Nature Genetics*, 24 (3), 227–35. https://doi: 10.1038/73432.

Rousseaux, Sophie, Alexandra Debernardi, Baptiste Jacquiau, Anne-Laure Vitte, Aurélien Vesin, Hélène Nagy-Mignotte, Denis Moro-Sibilot, et al. 2013. "Ectopic Activation of Germline and Placental Genes Identifies Aggressive Metastasis-Prone Lung Cancers." *Science Translational Medicine*, 5 (186), 186ra66-186ra66. https://doi.org/10.1126/scitranslmed.3005723.

Roychowdhury, Sameek, and Arul M. Chinnaiyan. 2016. "Translating Cancer Genomes and Transcriptomes for Precision Oncology." *CA: A Cancer Journal for Clinicians*, 66 (1), 75–88. https://doi.org/10.3322/caac.21329.

Sallinen, H., Janhonen, S., Pölönen, P., Niskanen, H., Liu, O. H., Kivelä, A., Hartikainen, J. M. et al. 2019. "Comparative Transcriptome Analysis of Matched Primary and Distant Metastatic Ovarian Carcinoma." *BMC Cancer*, 19 (1), 1121. https://doi.org/10.1186/s12885-019-6339-0.

Schaaf, Marco B., Abhishek D. Garg, and Patrizia Agostinis. 2018. "Defining the Role of the Tumor Vasculature in Antitumor Immunity and Immunotherapy." *Cell Death & Disease*, 9 (2), 115. https://doi.org/10.1038/s41419-017-0061-0.

Schwartzberg, Lee, Edward S. Kim, David Liu, and Deborah Schrag. 2017. "Precision Oncology: Who, How, What, When, and When Not?" *American Society of Clinical Oncology Educational Book. American Society of Clinical Oncology. Annual Meeting*, 37, 160–69. https://doi.org/10.1200/EDBK_174176.

Shalem, Ophir, Neville E. Sanjana, and Feng Zhang. 2015. "High-Throughput Functional Genomics Using CRISPR-Cas9." *Nature Reviews. Genetics*, 16 (5), 299–311. https://doi.org/10.1038/nrg3899.

Shaw, Alice T., and Jeffrey A. Engelman. 2014. "Ceritinib in ALK-Rearranged Non-Small-Cell Lung Cancer." *The New England Journal of Medicine*, 370 (26), 2537–39. https://doi.org/10.1056/NEJMc1404894.

Shaw, Alice T., Dong-Wan Kim, Kazuhiko Nakagawa, Takashi Seto, Lucio Crinó, Myung-Ju Ahn, Tommaso De Pas, et al. 2013. "Crizotinib versus Chemotherapy in Advanced ALK-Positive Lung Cancer." *The New England Journal of Medicine*, 368 (25), 2385–94. https://doi.org/10.1056/NEJMoa1214886.

Sheng, Kuanwei, Wenjian Cao, Yichi Niu, Qing Deng, and Chenghang Zong. 2017. "Effective Detection of Variation in Single-Cell Transcriptomes Using MATQ-Seq." *Nature Methods*, 14 (3), 267–70. https://doi.org/10.1038/nmeth.4145.

Shim, Gayong, Su-Eun Han, Yong-Hee Yu, Sangbin Lee, Han Young Lee, Kwangmeyung Kim, Ick Chan Kwon, et al. 2011. "Trilysinoyl Oleylamide-Based Cationic Liposomes for Systemic Co-Delivery of SiRNA and an Anticancer Drug." *Journal of Controlled Release: Official Journal of the Controlled Release Society*, 155 (1), 60–66. https://doi.org/10.1016/j.jconrel.2010.10.017.

Shoemaker, Robert H. 2006. "The NCI60 Human Tumour Cell Line Anticancer Drug Screen." *Nature Reviews. Cancer*, 6 (10), 813–23. https://doi.org/10.1038/nrc1951.

Shukla, S., Johnson, J. M., Curry, J. M., Kim, Y., Argiris, A., Luginbuhl, A.,Harshyne, L., and Rodeck, U. 2019. "Transcriptome Signatures of Treatment Responses in a Preoperative Window of Opportunity Trial of Nivolumab and Tadalafil in Resectable Squamous Cell Carcinoma of the Head and Neck." *Annals of Oncology*, 30 (October), v459. https://doi.org/10.1093/annonc/mdz252.023.

Stewart, Paul A, Eric A Welsh, Robbert JC Slebos, Bin Fang, Victoria Izumi, Matthew Chambers, Guolin Zhang, Ling Cen, Fredrik Pettersson, and Yonghong Zhang. 2019. "Proteogenomic Landscape of Squamous Cell Lung Cancer." *Nature Communications*, 10 (1), 1–17.

Stransky, Nicolas, Ethan Cerami, Stefanie Schalm, Joseph L. Kim, and Christoph Lengauer. 2014. "The Landscape of Kinase Fusions in Cancer." *Nature Communications*, 5 (September), 4846. https://doi.org/10.1038/ncomms5846.

Sung, Hyuna, Jacques Ferlay, Rebecca L. Siegel, Mathieu Laversanne, Isabelle Soerjomataram, Ahmedin Jemal, and Freddie Bray. 2021. "Global Cancer Statistics 2020: GLOBOCAN Estimates of Incidence and Mortality Worldwide for 36 Cancers in 185 Countries." *CA: A Cancer Journal for Clinicians*.,71 (3), 209-249. https://doi.org/10.3322/caac.21660.

Tang, Fuchou, CatalinBarbacioru, Yangzhou Wang, Ellen Nordman, Clarence Lee, Nanlan Xu, Xiaohui Wang, et al. 2009. "MRNA-Seq Whole-Transcriptome Analysis of a Single Cell." *Nature Methods* 6 (5): 377–82. https://doi.org/10.1038/nmeth.1315.

Tomczak, Katarzyna, PatrycjaCzerwińska, and Maciej Wiznerowicz. 2015. "The Cancer Genome Atlas (TCGA): An Immeasurable Source of Knowledge." *Contemporary Oncology (Poznan, Poland)*, 19 (1A), A68-77. https://doi.org/10.5114/wo.2014.47136.

Topalian, Suzanne L., Janis M. Taube, Robert A. Anders, and Drew M. Pardoll. 2016. "Mechanism-Driven Biomarkers to Guide Immune Checkpoint Blockade in Cancer Therapy." *Nature Reviews. Cancer*, 16 (5), 275–87. https://doi.org/10.1038/nrc.2016.36.

Tzelepis, Konstantinos, Hiroko Koike-Yusa, Etienne De Braekeleer, Yilong Li, Emmanouil Metzakopian, Oliver M. Dovey, Annalisa Mupo, et al. 2016. "A CRISPR Dropout Screen Identifies Genetic Vulnerabilities and Therapeutic Targets in Acute Myeloid Leukemia." *Cell Reports*, 17 (4), 1193–1205. https://doi.org/10.1016/j.celrep.2016.09.079.

Uhlen, Mathias, Cheng Zhang, Sunjae Lee, Evelina Sjöstedt, Linn Fagerberg, Gholamreza Bidkhori, Rui Benfeitas, et al. 2017. "A Pathology Atlas of the Human Cancer Transcriptome." *Science (New York, N.Y.)*, 357 (6352), https://doi.org/10.1126/science.aan2507.

Urban-Klein, B., Werth, S., Abuharbeid, S., Czubayko, F., and Aigner, A.2005. "RNAi-Mediated Gene-Targeting through Systemic Application of Polyethylenimine (PEI)-Complexed SiRNA *in Vivo*." *Gene Therapy*, 12 (5), 461–66. https://doi.org/10.1038/sj.gt.3302425.

Varadhachary, Gauri R., Yael Spector, James L. Abbruzzese, Shai Rosenwald, Huamin Wang, Ranit Aharonov, Heather R. Carlson, et al. 2011. "Prospective Gene Signature Study Using MicroRNA to Identify the Tissue of Origin in Patients with Carcinoma of Unknown Primary." *Clinical Cancer Research: An Official Journal of the American Association for Cancer Research*, 17 (12), 4063–70. https://doi.org/10.1158/1078-0432.CCR-10-2599.

Vempati, Uma D., Caty Chung, Chris Mader, Amar Koleti, Nakul Datar, Dušica Vidović, David Wrobel, et al. 2014. "Metadata Standard and Data Exchange Specifications to Describe, Model, and Integrate Complex and Diverse High-Throughput Screening Data from the Library of Integrated Network-Based Cellular Signatures (LINCS)." *Journal of Biomolecular Screening*, 19 (5), 803–16. https://doi.org/10.1177/1087057114522514.

Wang, Huajing, and William Sun. 2017. "CRISPR-Mediated Targeting of HER2 Inhibits Cell Proliferation through a Dominant Negative Mutation." *Cancer Letters*, 385 (January), 137–43. https://doi.org/10.1016/j.canlet.2016.10.033.

Wang, Maonan, Jingzhou Zhao, Lishen Zhang, Fang Wei, Yu Lian, Yingfeng Wu, Zhaojian Gong, et al. 2017. "Role of Tumor Microenvironment in Tumorigenesis." *Journal of Cancer*, 8 (5), 761–73. https://doi.org/10.7150/jca.17648.

Wang, Maonan, Jingzhou Zhao, Lishen Zhang, Fang Wei, Yu Lian, Yingfeng Wu, Zhaojian Gong, et al. 2017. "Role of Tumor Microenvironment in Tumorigenesis." *Journal of Cancer*, 8 (5), 761–73. https://doi.org/10.7150/jca.17648.

Wang, Xin Wei, and Snorri S. Thorgeirsson. 2009. "Transcriptome Analysis of Liver Cancer: Ready for the Clinic?" *Journal of Hepatology*, 50 (5), 1062–64. https://doi.org/10.1016/j.jhep.2009.02.007.

Wu, Yi-Mi, Fengyun Su, Shanker Kalyana-Sundaram, Nickolay Khazanov, Bushra Ateeq, Xuhong Cao, Robert J. Lonigro, et al. 2013. "Identification of Targetable FGFR Gene Fusions in Diverse Cancers." *Cancer Discovery*, 3 (6), 636–47. https://doi.org/10.1158/2159-8290.CD-13-0050.

Yao, Yao, Zhongyi Yan, SenlinLian, Liangnian Wei, Chao Zhou, Dongju Feng, Yuan Zhang, Jianrong Yang, Ming Li, and Yun Chen. 2020. "Prognostic Value of Novel Immune-Related Genomic Biomarkers Identified in Head and Neck Squamous Cell Carcinoma." *Journal for ImmunoTherapy of Cancer*, 8 (2), e000444. https://doi.org/10.1136/jitc-2019-000444.

Zevallos, Jose P., Angela L. Mazul, Vonn Walter, and David Neil Hayes. 2019. "Gene Expression Subtype Predicts Nodal Metastasis and Survival in Human Papillomavirus–Negative Head and Neck Cancer." *The Laryngoscope*, 129 (1), 154–61. https://doi.org/10.1002/lary.27340.

Zhan, Tianzuo, NiklasRindtorff, Johannes Betge, Matthias P. Ebert, and Michael Boutros. 2019. "CRISPR/Cas9 for Cancer Research and Therapy." *Seminars in Cancer Biology*, 55 (April), 106–19. https://doi.org/10.1016/j.semcancer.2018.04.001.

Zhang, Shumei, Yihan Wang, Yue Gu, Jiang Zhu, Ce Ci, Zhongfu Guo, Chuangeng Chen, et al. 2018. "Specific Breast Cancer Prognosis-subtype Distinctions Based on DNA Methylation Patterns." *Molecular Oncology*, 12 (7), 1047–60. https://doi.org/10.1002/1878-0261.12309.

Zhen, Shuai, Ling Hua, Y. Takahashi, S. Narita, Yun-Hui Liu, and Yan Li. 2014. "*In Vitro* and *in Vivo* Growth Suppression of Human Papillomavirus 16-Positive Cervical Cancer Cells by CRISPR/Cas9." *Biochemical and Biophysical Research Communications*, 450 (4), 1422–26. https://doi.org/10.1016/j.bbrc.2014.07.014.

Part III. Transcriptomics in Contemporary Sciences

Chapter 6

Role of Transcriptomics in Understanding the Dynamics of Host-Pathogen Interaction

Sekhu Ansari[1,*], Kasturi Roy[2]
and Dharmendra Nath Bhatt[3]
[1]Division of Pathology, Cincinnati Children's Hospital, Cincinnati, Ohio, USA
[2]Department of Molecular Biology and Biotechnology, University of Kalyani, Nadia, West Bengal, India
[3]National Institute of Plant Genome Research,
Jawaharlal Nehru University Campus, Aruna Asaf Ali Road, New Delhi, India

Abstract

Host cells are exposed to a plethora of microbial pathogens such as viruses, bacteria, fungi, protozoans. Each microbial interaction with the host has a unique dynamics/interplay of events required for the survival, adaptation, and evolution of the pathogens as well as the host. Understanding these highly complex host-pathogen interactions are a requisite for a better understanding and prevention of infectious diseases. Continuous technological revolutions have refined our ability in flawless study of gene expression. The architecture of the gene expression network of these interactions could be decoded with high-throughput RNA sequencing. Transcriptomics technologies has become a powerful indispensable tool for the exploration of the association of a single-cell bacterium to the multicellular eukaryotic organism. Here, we describe the recent advancements and developments in transcriptomic approaches, involved in the exploration of host-pathogen interactions. Single-cell RNAseq and dual RNAseq are the most common and innovative

---

[*]Correspondence Author's Email: shekhuansar@gmail.com.

In: Transcriptomics and their Importance
Editors: Sarah Naiyer and. Khalid Hussain Bhat
ISBN: 979-8-88697-044-9

approaches that have given us insights in the cell-to-cell interaction studies. Transcriptomic studies in infectious diseases also helped us identify the potential unique set of transcripts to be used as biomarkers for the disease. Moreover, the dual RNAseq has been successfully implemented in the study of various microbial pathogens that causes infection in humans and cereal plant cells such as *Mycobacterium tuberculosis, Salmonella* species, *Xanthomonas oryzae, etc* Further advancements in transcriptomic analysis require the integration of transcriptomics with the proteomics, metabolomics and genomics data along with the inclusion of other scientific disciplines like mathematical modelling. In conclusion, transcriptomics analysis of host-pathogen interactions reveals the dynamics of the molecular changes and helps in a better understanding of the disease and ways to combat its progression.

Keywords: parasitism, host-pathogen interactions, infectious disease, single-cell RNAseq, biomarkers

1. Introduction

Parasitism is among one of the most prevalent biological interactions on earth. Parasites depend upon their host for nutrient acquisition (Zuzarte-Luís and Mota 2018). Once the host cell encounters the parasite, the arms race between them starts. This molecular race is a balance between the virulence of the pathogen and the host defence. In this millions of years old race, the host tries to get the better of pathogen through its robust immune system while the parasite attempts to establish the infection by evading the immune response of the host. Several hosts have evolved the strategies to triumph the molecular war with the pathogen. However, several pathogens have also devised complex strategies to infect the host and complete their life cycle. In the course of human history, both plant and human pathogens have created havoc and have been responsible for the loss of millions of human lives. Pathogens remain a significant threat to their host. The emergence of drug resistance in human pathogens warrants the attention of the new drug targets. Similarly, plant pathogens resistant to pesticides are also emerging. So, the development of new drug targets or drugs against pathogens is the need of an hour. During the course of host-pathogen interaction, there are dynamic changes in the expression of genes of both host and pathogens. There is a need to study these gene expression dynamics. A comprehensive understanding of these dynamics will improve our ability to develop new strategies against pathogens. To gain a clear understanding of the host-pathogen interaction, the dynamics of gene expression need to be studied. In recent years, genome-wide omics analysis has emerged as an important tool in biological research for studying the biochemical pathways involved in the host-pathogen interaction.

The emergence of omics studies has vastly improved our understanding of the host-pathogen interaction. Transcriptomics is one such technique that is quickly evolving into omics studies. Transcriptomics focuses on the analysis of gene expression and provides global genome-wide information. Transcriptome analysis along with the combination of other omics studies like metabolomics and proteomics of both host and pathogens during and after infection has helped to decipher the insights of the host-pathogen interactions. Here in this chapter, we present the advancements and the current state of transcriptomics and its applications in the field of host-pathogen interaction.

1.1. Host-Pathogen Interaction

Over half of the life on the earth is parasitic in nature. Parasitic diseases have several times caused havoc on earth. Human diseases like the plague was responsible for the loss of significant human life (Glatter and Finkelman 2021). Similarly, several plant diseases like the blight disease of potatoes were responsible for the loss of ample amount of crops which eventually was the cause of the loss of human life (Goss et al. 2014). Pathogenic diseases in both plants and animals are a result of the intimate molecular relationship between the pathogen and its host. Hence, the term host-pathogen interaction is widely used for such an interactive relationship of the pathogen with its host. Host-pathogen interaction starts with the infection of the host by the pathogen. After the infection, host-pathogen interaction involves two more stages: replication of the pathogen inside the host and the host response to eliminate the pathogen (Sen, Nayak, and De 2016).

1.1.1. Infection of Host

The first step in the invasion of the host cell is to recognize and adhere to the host cell. Pathogens recognize their host cell through their molecular structures called pathogen associated molecular patterns (PAMPs) (Mogensen 2009). PAMPs interact with the host pattern recognition receptors (HPRRs), like toll-like receptors in animals and receptor kinases in plants, to adhere to the host cell (Mogensen 2009; Chetouhi et al. 2015). During and after the adhesion, pathogens usually secrete effector molecules which promote their invasion and compatible interaction with the host. Many Gram-negative bacteria secrete different toxins which are essential for their pathogenesis. Pathogens like *Mycobacterium species, H. pylori,* etc. secrete urease which helps in raising the pH of the environment thereby helping them in generating a conducive environment for their survival (Clemens, Lee, and Horwitz 1995). Similarly, plant pathogens like *Magnaporthe oryzae, Alternaria alternate,* etc. secrete cell wall degrading enzymes (CWDEs) (Bhatt et al. 2020; Ma et al. 2019). Enzymes like pectate lyases, polygalacturonase, cutinase, β-glucosidase, etc. are among a few of the cell wall degrading enzymes secreted by plant pathogens for their invasion (Ma et al. 2019; Bhatt et al. 2020).

1.1.2. Replication of Pathogen

Before starting its replication inside the host, pathogens must evade the host defence. Hence, after the invasion of the host, the pathogen shifts its focus on evading the host's defence. Pathogens usually interfere with the essential pathways of the host. Animal pathogens must evade the macrophage. For this, different pathogens have different strategies. Pathogens can either inhibit the movement of the macrophages by the enzymatic degradation of the actin, disrupt the activation of the macrophages or enter the phagolysosome and prevent the maturation of the phagosomes (Eisenreich et al. 2010). Similarly, plant pathogens have also evolved strategies for countering the host defence. Plant pathogens counter the host defence by reprogramming the immune signal transduction pathways of the host or by adapting to the immune microenvironment of the host (Chetouhi et al. 2015). Pathogens like *Yersinia pestis* secrete effector proteins or produce secondary metabolites to modulate the host immune system. Both plant and animal pathogens are often seen to exploit the host cell for their nutritional requirements during replication. Pathogens also modulate their host interaction through micro RNAs. After the successful infection of a host, pathogens replicate within the host. Intracellular pathogens like *Mycobacteriaand Leishmania* have evolved several strategies for their proliferation inside the host cell (Ansari et al. 2021). Many pathogens hijack the host cell and use nutrients from them. Viruses replicate by integrating their genetic material into the host genome (Weitzman and Fradet-Turcotte 2018). Sometimes pathogens like *C. albicans* form the biofilm for their replication (McCall et al. 2019).

1.1.3. Host Response

Host cells try to restrict the replication of the pathogens. Host cells utilize a diverse range of strategies to prevent the manifestation of the pathogen. The human body launches a strong immune response upon the pathogenic infection either by activating the dendritic cells, altering the expression of several defence mechanism genes, autophagy, or by nutrient limitation (Martin-Gayo and Yu 2019)(Sen, Nayak, and De 2016). In humans, cells at the site of infection recruit dendritic cells. Upon activation, dendritic cells release several chemokines. During infection, several genes responsible for the activation of immunologic response are also expressed. Autophagy is another strategy by the host to restrict the proliferation of parasites. In some intracellular pathogenic infections, human macrophages restrict the availability of nutrients. For *e.g.,* in several pathogenic infections supplies of iron is restricted by the host cells. Like animal cells, plant cells have also devised strategies against parasites. Plant cells can thicken their cell wall, acidify their apoplast or induce the transcription of the defense mechanism genes (Chetouhi et al. 2015). Some plants show hypersensitive responses to parasites. During the hypersensitive response, plant cells restrict the pathogenic infection by altering the ion flux, reactive oxygen species generation, or by secreting the antimicrobial compounds (González-Lamothe et al. 2009) (Huang et al. 2019). Host-pathogen interactions are highly complex, involving large fluctuating changes in gene expression. The major aim of the study of the host-pathogen interaction is to understand the way pathogens and host disrupt each other's biochemical pathways. An unbiased global understanding of the transcriptome of both host and pathogen provides new insight into the various key metabolic pathways altered during their interaction. This understanding of the host pathogen interaction will help in the discovery of novel anti-parasitic targets and compounds.

2. Various Advancements in Transcriptomics Methods for Microbial Studies

2.1. Evolution in Transcriptomics in Resolving Pathogen Interactions

In the past few decades, transcriptomics technologies have been highly developed and become an essential part of biological research. After reporting the first ribonucleotide sequence in 1964 which is 77-nucleotides yeast alanine t-RNA, the sequencing technology evolved (Kresge, Simoni, and Hill 2006). After the invention of microarray technology by Patrick O. Brown in 1981, sequencing technology revolutionized and shifted from the postgenomic era to the functional genomics era (Brown and Botstein 1999). The microarray analyses defined a set of gene expressions by the hybridization to their complementary probes (Hoheisel 2006). The term, "transcriptome" was first used in 1990 for bulk gene expression analysis (Lowe et al. 2017).

Whole genome sequencing of microbial pathogens and their hosts is revolutionizing the new methodology to explore the host-pathogen interactions (Quainoo et al. 2017). DNA microarrays were the first sophisticated technique used in the study of predefined transcripts/gene expression involved in host-pathogen interaction (Cummings and Relman 2000). In 1998, Saizieu and his colleagues developed a chip containing 206 probes representing the genes of the *Hemophilus influenzae* and *Streptococcus pneumonia*. By this, they identified 11 genes differentially expressed during stationary and exponential phases (Saizieu et al. 1998).

Different microbial organisms are classified according to their interaction with the host. The host-microbe interaction may be beneficial to the host as in the case of mutualism, or it may not as in the case of parasitism. Pathogenic bacteria, protozoans reside inside the host cell. Few bacteria such as *Rickettsia* species live in the cytoplasm and rarely multiply inside the nucleus (Mansueto et al. 2012). Several bacteria and protozoan e.g., *Chlamydia* species, *Coxiella burnetii*, *Leishmania* and *Trypanosoma* species use the host's phagosome and phagolysosome for survival and escape from the host immune response (Borawski et al., 2020) (Wong et al. 2019) (Radwanska et al. 2018) (Ansari et al. 2021). Past generation molecular biology techniques could explore the limited number of genes. In contrast to this, microarrays allow the study of a large number of genes expressed during the host-pathogen interactions (Leroy and Raoult 2010). In early transcriptomics, a Sanger sequencing-based serial analysis of gene expression (SAGE) method was developed which gave an entire transcript expression profile of cells for specific conditions and it also helps in the identification of the new genes (Kukurba and Montgomery 2015). Over the last few years, another technique RNAseq has been developed which is high-throughput sequencing technology. In comparison to the Sanger sequencing and microarray, RNAseq is more sensitive, highly dynamic, and provides a whole landscape of expressed identified and unidentified gene patterns. Moreover, the sensitive RNAseq can identify different alternative spliced transcripts and different allele-specific expressions. In addition to messenger RNA (mRNA) transcripts, pre-mRNA, and noncoding RNA (microRNA and long noncoding RNA) populations can be analysed (Wang, Gerstein, and Snyder 2009) (Heather and Chain 2016).

2.2. Advancement in Transcriptomics in Resolving Pathogen Interactions

Various microbial pathogens such as viruses and bacteria evolve themselves which leads to a very dreadful pandemic. In 1918, the well-known maleficent Spanish influenza claimed almost 50 million people's life (Scarpa et al. 2020). Today we are facing the COVID-19 pandemic which was first reported in December 2019, in Wuhan, China (Izda, Jeffries, and Sawalha 2021). To combat the rising threats with the diverse microbial pathogens (e.g., viruses, bacteria, protozoa, fungi, and worms), it is requisite to understand the host-microbe interactions. In the current scenario, paradigms of host-microbe interaction are shifted and redefined (Sen, Nayak, and De 2016). The new-age research reveals that pathogen and host interaction does not always lead to disease. Inside the human host, *Mycobacterium tuberculosis* resides in a dormant stage without causing any disease symptoms. (Gengenbacher and Kaufmann 2012). Commensal, carrier state, hemibiotrophic, and opportunist are the new terminology that defines more behaviour and nature of host-pathogen interactions (Casadevall and Pirofski 2000). Nowadays, highly advanced transcriptomics has become a very powerful tool to understand the numerous mechanism and complex associations between the host and its microbial pathogens. Single-cell RNA sequencing (scRNA-seq) and dual RNA-seq are the two developed high throughput sequencing technologies that can help in the investigation of host-pathogen interactions.

2.2.1. Single-Cell RNA Sequencing (scRNA-Seq)

With the passing of time, the highly complicated host-pathogen association is co-evolved due to the various intrinsic and extrinsic factors. The recent and advanced single-cell RNA sequencing (scRNA-seq) provides usability to explore the fundamental mechanism of successful host-pathogen interactions and the host's innate and adaptive immune response against microbial infections (Chattopadhyay, Roederer, and Bolton 2018). The biological techniques and *in silico* biology revolution add insight into deep RNA sequencing ability for a single cell expression analysis. James Eberwine and his colleague pioneered the concept of single cell RNA sequencing and demonstrate the expression profiling of a single cell neuron isolated from the rat's hippocampus (Eberwine et al. 1992). In 2009, Fuchou Tang and his research team for the first time reported the next generation sequencing based transcript expression profile of a single cell blastomere for comparative analysis with microarray expression analysis (Tang et al. 2009).

2.2.1.1. Single-Cell Segregation Techniques

The separation of a single cell from the heterogeneous cell population is the first and crucial step in the scRNA-seq analysis. Total RNA isolated from the whole organism or the tissue give an average response and less information because individual cell type has different gene expression profile (Tang et al. 2009) (Hwang, Lee, and Bang 2018). Identification of differential cell to cell RNA transcript and protein expression profile is crucial and can provide the therapeutic solution to cancer, neurodegenerative disorders, and other malicious diseases.

In traditional methods, single cells are isolated by the dilution method. In the principle of the dilution method, higher dilution of densely populated heterogeneous cells increases the higher chance of getting individual cells in a single drop (Gross et al. 2015). In another classical method, microscope-guided pipettes have been used to isolate single cell (Hwang, Lee, and Bang 2018). However, these classical methods are time-consuming and laborious. Nowadays, a recently developed flow-activated cell sorting (FACS) has become a well-established method for individual cell segregation. In this technique, a fluorescent monoclonal antibody is used to tag a specific cell, which allows to isolate it from the heterogeneous cells population by flow-activated cell sorting (McKinnon 2018). Laser capture microdissection (LCM) is another highly advanced technique to segregate the cells and cell compartments from the solid tissue. A microscope-guided and directly observed desired cells are cut off with a laser beam. Past few years microfluidic or lab-on-a-chip devices have been used in the isolation of single cells and analysis. Due to its small sample requirement and low cost, it gained popularity. The nanoliter-sized volume of tissue samples is required which also helps in the reduction of contamination. Initially, microfluidics was used for the small scale biochemical assays for DNA and protein analysis (Hu et al. 2016). Now its rapid expansion and evolution help in the analysis of the transcript expression profile of single cells such as bacteria.

2.2.1.2. Sequencing Platform Selection and scRNA-Seq Library Preparation

At present time various single-cell sequencing platforms are available like 10x genomics chromium, Nadia (Dolomite Bio), Illumina Bio-Rad ddseq single-cell isolator, ICELL8 single-cell system (Takara), Fludigm C1, BD Rhapsody single-cell analysis system (BD), etc (Valihrach, Androvic, and Kubista 2018). Among them, 10x genomics chromium is very common and widely used. Choosing the perfect and suitable platform is essential for the success of the high-throughput scRNA-seq.

Before using the sequencer platform, the common steps of scRNAseq libraries preparation includes RNA isolation from cell, first-strand cDNA synthesis by using reverse transcriptase enzyme and cDNA amplification. The second strands are further generated by either poly(A) tailing or by a template switching mechanism (Tang et al. 2009) (Hwang, Lee, and Bang 2018a). Later the synthesised DNA is amplified by using typical PCR or in vitro transcription. Further, Smart-seq2 generates a full-length transcripts sequence which is also suitable for alternative splicing and allele-specific expression discovery (Mamanova et al. 2021). On average each cell has 10^5 to 10^6 messenger RNA and more than ten thousand proteins. For this recently researcher use the unique molecular identifier or barcodes (4-8 size random base pair) system during the synthesis of first-strand cDNA (Sena et al. 2018). These unique coding systems strengthen and highly improved the quantification and sequencing data quality.

2.3. Dual-RNAseq

The first transcript expression analysis for the host and its microbial pathogen was performed with the development of cDNA microarrays. The gene expression profile of various bacterial pathogens such as *Vibrio cholerae*, *Chlamydia trachomatis*, *Borrelia burgdorferi*, and *Salmonella enterica* with their specific hosts was first studied with microarrays that reveal the blueprint of the infection strategies of these pathogens (Westermann, Gorski, and Vogel 2012). After the development of probe independent RNA sequencing approach that revolutionized transcriptomics by producing high quality and high quantity sequencing data (Westermann, Gorski, and Vogel 2012). Taking another leap in the next-generation sequencing for the host-pathogen interaction study, the concept of Dual-RNAseq has emerged. In this new methodology, the combined total RNA isolated from the host and its interacting microbial partner is simultaneously sequenced and analysed. Dual RNA-seq was initially applied for the viral, fungal, and parasite infection which shows the transcriptome architecture similarity with its host. The utmost advantage of dual-RNAseq is the high throughput expression analysis of transcripts expressed during the interaction of host and microbial pathogen (Nuss et al. 2017).

Eukaryotic host cells have a more complex genome than their microbial pathogen. The size of the human genome is 3,000 Mb which is 600 times greater in size than *E. coli* 5 Mb in size (Fields et al. 1994) (Kaper, Nataro, and Mobley 2004). Approximately 21,306 protein-coding genes are reported in humans, and 4,200 protein-coding genes were reported in bacterial cells (Willyard 2018) (Moreno-Campuzano, Janga, and Pérez-Rueda 2006). 10-20 pg total RNA is present in eukaryotic cells and 0.1 pg total RNA is present in a bacterial cell (Westermann, Gorski, and Vogel 2012). After the bacterial infection, each host cell contains more than one bacterial cell which increases the intracellular copy number for the microbial pathogen. In a mammalian host, on average each cell is infected with ten bacteria. Because of that, the relative difference of bacterial total RNA content will need to decrease to about 10-20-fold in most bacterial infections. On the other side due to the small genome size of prokaryotic microbes, approximately 10 to 20 times more reads are required for the dual RNA-seq (Westermann, Barquist, and Vogel 2017) (Westermann, Gorski, and Vogel 2012). In eukaryotic cell to cell interactions, the expressed total RNA amount is similar. Still, a high number of reads is required, due to the small genome size of eukaryotic microbes. Different host-microbe interactions are awaiting to explore with firmly well-established dual RNA-seq analysis. The following few steps are crucial in dual RNA-seq analysis experiments designing and analysis.

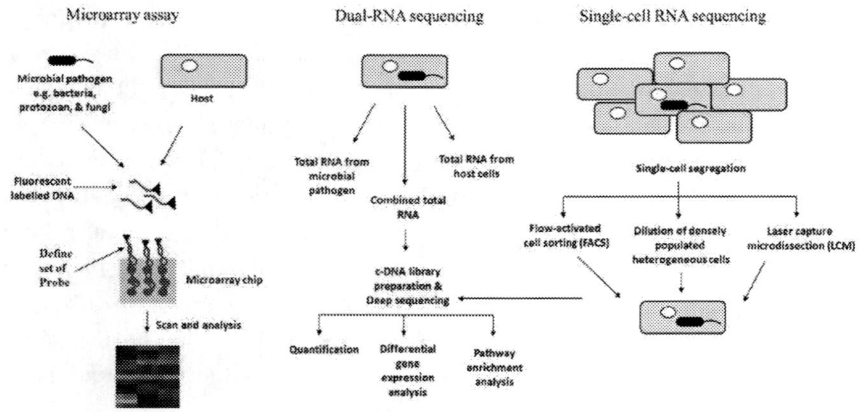

Figure 1. A comparative concept of transcriptomics in the study of host-pathogen interaction.

2.3.1. Total RNA Isolation

A sufficient threashold of infected cells is required for the study. In most studies, at least 10,000 infected cells are required. The selection and separation of infected cells from the non-infected cells are critical and essential for the dual RNA-seq analysis (LaMonte et al. 2019). In many studies, green fluorescent protein expressing microbial cells is used to distinguish between infected and non-infected host cells. Laser capture microdissection and FACS are frequently used to separate the infected cells (Ambriz-Aviña, Contreras-Garduño, and Pedraza-Reyes 2014). A similar approach was used in the infection study of the *Salmonella* species and *Streptococcus pneumonia* (Haidinger et al. 2001). Once enough individual infected cells are collected, the cells must be lysed mechanically or chemically to extract the total RNA. Thereafter, isolated total RNA needs to purify to remove the genomic DNA contamination. Before RNA sequencing the comparative concentration of microbes and their host needs to be analysed with quantitative real-time PCR (qRT-PCR). This will define how much read depth is required for sequencing and any amendment required in the protocols such as increasing the number of infected cells.

3. Application of Transcriptomics in Host-Pathogen Interactions

Nowadays transcriptomics is widely used to understand the host-pathogen interaction. Both coding and non-coding RNAs are used in this study to understand the disease development and defense mechanism inside the host cell during infection (Sudhagar, Kumar, and El-Matbouli 2018). It also reveals the strategies for infection and survival of pathogens during infections. (Westermann, Barquist, and Vogel 2017). These findings will help to strategize the therapeutic processes and drug development. Several transcriptomics technologies like microarray, SAGE, cap analysis gene expression (CAGE), RNA-Seq, dual RNA-Seq, bulk RNA-Seq, scRNA-Seq have evolved over time to understand host-pathogen interactions (Lee et al. 2018). Some of the host-pathogen interaction studies in animal and plant hosts by transcriptomic analysis from already published data are given in the tables.

3.1. Findings from the Transcriptomic Analysis from Host Pathogen Interaction Studies

3.1.1. Transcriptomics Reveals the Strategies for Infection and Survival of Pathogens during Infections

Transcriptomic analysis helps to find out the important pathogen genes involved in host infections and reveals the strategies for immune evasion and survival inside the host cells. *Magnaporthe oryzae* infection in rice revealed putative secretory proteins mediated attack of the host cells at the initial infection stage (Kawahara et al. 2012). *X. fragariae,* infection in strawberry revealed upregulation of genes involved in host interaction, recognition, pathogenesis, and suppression of the host innate immune response (T3E genes) (Gétaz et al. 2020). A previous study also revealed the involvement of *P. cinnamomi* gene (crinkler effector) CRN1 in the suppression of the host defense mechanism in *E. nitens* (Meyer et al. 2016). *Orientia tsutsugamushi* (Karp strain) infection in HUVECs revealed breakage of noncoding RNAs and transcribed operons through frequent recombination and transposition by RAGE and ICE (Mika-Gospodorz et al. 2020). The study also revealed the upregulation of RAGE genes by antisense transcription (Mika-Gospodorz et al. 2020). Differential expressions of genes in *C. gattii* lineage were found VGII compared to non-VGII lineages like capsule synthesis, capsule attachment, and ergosterol biosynthesis genes during mouse BMDMs infection (Farrer et al. 2018). VGII lineage is the reason for most *C. gattii* infections (Farrer et al. 2018). It also revealed upregulation of ergosterol biosynthesis and switch of the expression from LAC1 to LAC2 involved in the protection of VGII lineage against host defense (Farrer et al. 2018). *T. gondii* infection in mouse PK-15 cells revealed upregulation of cell invasion genes (DHFR-TS), PV formation genes, PCNA gene, and reprogramming of pathogen metabolic processes to alter the microenvironment of PK-15 cell for the establishment of infection in porcine (Zhou et al. 2016). Induction of pro-inflammatory M1 polarization to evade host immune system was found in mice MΦs against non-growing intracellular *Salmonella* and induction of anti-inflammatory response against intracellular growing *Salmonella* to overcome host defense by changing polarization state (M2) of mice MΦs was also found in this study (Saliba et al. 2016). Increased expression of ADI operon to survive in septic and hypoxic A/J mice cell were found in *S. aureus* during infection (Thänert et al. 2017). This study also revealed increased expression of cytotoxins and extracellular proteases in *S. aureus* to survive in resistant C57BL/6 mice (Thänert et al. 2017).

3.1.2. Transcriptomics Reveals Disease Development and Defense Mechanism of Host Cells against Pathogen

Transcriptomic analysis of pathogens infected host cells helps to understand the disease development and defense mechanism of host cells during infections. *M. oryzae* infection in *Oryzae sativ*a revealed the involvement of PR genes in host the defense mechanism (Kawahara et al. 2012). SMV infection in soybean plants revealed upregulation of salicylic acid (SA), jasmonic acid (JA), and ethylene (ETH) pathways and PR genes in the presence of light leading to resistance against the pathogen (Kawahara et al. 2012; L. Zhang et al. 2019). *X. fragariae* infection in strawberry revealed resistance against pathogen by limiting nutrition and inducing PAMP receptors at the early infection stage in the host cell and by upregulating pathogen recognition in later a stage (Gétaz et al. 2020). *P. cinnamomi* infection in *E. nitens* revealed the involvement of PR genes in the defense mechanism via SA and JA signaling pathways (Meyer et al. 2016). PMev infection in *Carica papaya* revealed involvement of SA signaling at the early stage in the prevention of sticky disease syndrome but possible involvement of NPR1-inhibitor candidate gene, NPR1- I/NIM1-I, UGTs, and ET in the prevention of SA-associated resistance against PMeV complex in a later stage (Madroñero et al. 2018). *F. verticillioides* infection in Sugar cane revealed the involvement of PR related genes like Unigene0030984, Unigene0015553, Unigene0012327, Unigene0044480, Unigene0008546 and Unigene0064857 in resistance mechanism of host cells (Z. Wang et al. 2019). Cereal cyst nematode infection in *Aegilops variabilis* revealed involvement of plant-pathogen phosphatidylinositol signaling system and ABC transporters pathways in CCN resistance (Xu et al. 2012). Involvement of PIF-like ORF1 and NBS-LRR type genes in the resistant mechanisms was revealed in rice against *Burkholderia glumae* (Magbanua et al. 2014). Involvement of SA, JA, and ET pathways in defense mechanism was revealed in wheat against Bgt infection but involvement of ABA signalling pathway in inhibition of these pathways was also revealed in susceptible interaction (J. M. Wang et al. 2012). The establishment of the immune system by chitin response and ethylene mediated signalling pathway was revealed in chrysanthemum against CMV, TSWV and PVX (Choi et al. 2015).

Type-I IFN signalling pathway was revealed in HUVEC against *O. tsutsugamushi* infection (Mika-Gospodorz et al. 2020). IL6 and IL33-NOS3-

FAS mediated strong pro-inflammatory responses in UT176, and Karp were also found respectively (Mika-Gospodorz et al. 2020). FosB/Jun/Egr1 regulatory proteins mediated host response were also found in mouse BMDMs against *C. gattii* infection (Farrer et al. 2018). At theearly stage of infection higher expression of immune-related genes associated with increased IFN-γ production were revealed in H5N8-infected mice which may play anti-viral immunity in the host cells (Park et al. 2015). Moderate expression of chemokines leading to moderate virus sensing and virus-mediated inflammation was later revealed in H5N8 infected mice but higher expression of chemokines leading to severe injury in the lung tissue was also revealed in H5N1 infected mice (Park et al. 2015). Induction of the TGF-β signalling pathway, p53 signalling pathway and rap1 signalling pathway, carcinogenesis and amoebiasis were also revealed in porcine PK-15 cells in response to *T. gondii* (Zhou et al. 2016). LPS-induced immune response and chemokines mediated inflammatory response were revealed in mice MΦs in response to *Salmonella* (Saliba et al. 2016). Induction of cholesterol biosynthesis to resist SARS-CoV-2 infection was observed in human alveolar basal epithelial carcinoma cells (Daniloski et al. 2021). Rapid induction of extracellular growth factors followed by upregulation of AP1 transcription factor complex (FOS and JUN) were found in human kidney cells (HEK293T) in response to Ebola virus infection (Wynne et al. 2017). The study also revealed upregulation of vascular development (angiogenesis), extracellular matrix organization, and coagulation by AP1 in the host cells (Wynne et al. 2017). *S. aureus* infection in A/J and C57BL/6 mice revealed cytokine-cytokine receptor interaction and chemokine signalling pathway mediated host response involved in the intense inflammatory response and pre-coagulation response in A/J mice leading to severe sepsis and death (Thänert et al. 2017). Induction of highly hypoxic conditions and arrest of bacterial multiplication were found as a defense mechanism in A/J mice and C57BL/6 mice against *S. aureus* (Thänert et al. 2017). The increased expression level of specific miRNAs which reflects apoptosis, proliferation (PI3K-Akt, AMPK, and ErbB pathways), and growth in sheep lung tissue were found against SRLV (Small Ruminant Lentiviruses) infection (Bilbao-Arribas et al. 2019). The study also revealed the possible involvement of oar-miR-21 mediated inflammation and proliferation in lesions formation of sheep lungs (Bilbao-Arribas et al. 2019).

Table 1. Overview of some animal-pathogen interaction studies using transcriptomic analysis from already published data

Host/pathogen	Method used	Findings in host	Findings in microbes	Reference
Human/ *Orientia tsutsuga-mushi* (Ot)	dual RNA-Seq of Ot infected umbilical endothelial cells	Pro-inflammatory chemokine and TLR3 pathway upregulated	DNA replication and metabolism were upregulated and surface protein encoding genes scaE, tsa56 and tsa22 are highly expressed	(Mika-Gospodorz et al. 2020)
Mouse/Cryptoco-ccus gattii	dual RNA- Seq of five *C. gattii* infected bone marrow derived macrophages in vitro and ex vivo at 1, 3, and 6 hpi.	Induction of FosB/Jun/Egr1 regulatory pathway	induction of oxidative-reduction functions and capsule attachment genes are downregulated	(Farrer et al. 2018)
Mice/ H5N8 & H5N1 influenza virus	dual RNA-Seq of H5N8 and H5N1 infected lungs of mice at 1,3 and 7 dpi	Chemokines and interferon stimulated genes upregulated in H5N1-infected mice. Activation of p38 MAPK, iNOS signalling in H5N8 and death receptor signalling in H5N1- infected mice.	HA, M, NA, NP, NS, PA, PB1 and PB2 transcripts level were low in H5N8 infected mice but HA, NS1 and NA transcript level were high in H5N8-infected mouse than H5N1-infected mouse.	(Park et al. 2015)
Porcine kidney (PK-15) cells/ Toxoplasma gondii	dual RNA- Seq from *T. gondii* infected kidney (PK-15) cells at 1, 3, 6, 9, 12, 18, and 24 hpi	Host immune genes (JUNB, CCL5, HSP90B1, FASN, TFDP1, MAP3K6, LAMC2, MALT1, RASSF5, ICAM-1, CCNE2, GADD45G, SFN, CISH_TV2, TGFB1, LAMA3, HSP70.2, GADD45B, ACKR3, and SRRM2) Upregulated and MAP3K3, IGFBP3, ITGB5, IGF1R, CACNA2D1, GUCY1B3, PRKC1, GUCY1A3, CSF1, MCSF_ALPHA, EPHA1, MCL1, FGFR1IIC, ERBB3, NOD1, THBS1, PLCD1, and CDC25B genes down regulated	Among the 68 KEGG pathways, DEGs were mostly involved in metabolic pathways, biosynthesis of secondary metabolites, pyrimidine metabolism and carbon metabolism. The altered metabolic pathways of *T. gondii* revealed host cell wall modification to easily infect the host	(Zhou et al. 2016)
Mice/*Salmonella typhimurium*	single cell RNA-Seq from invitro *S. typhimurium* infected bone marrow-derived	Pro-inflammatory factors like NF-κB, Tnf, Cxcl10, Il1b, NLRP3 and TLR2 similarly upregulated and DNA metabolism, cellular metabolic process in were down regulated		(Saliba et al. 2016)

Table 1. (Continued)

Host/pathogen	Method used	Findings in host	Findings in microbes	Reference
Human/SARS-CoV-2	single cell RNA-Seq of SARS-CoV-2 infected human alveolar basal epithelial carcinoma cells	Perturbation of RAB7A, PIK3C3, NPC1, CCDC22, ATP6V1A and ATP6AP1 genes were involved in upregulation of cholesterol synthesis, NPC1, ATP6V1F and ATP6V1G1 genes		(Daniloski et al. 2021)
Human/Ebolavirus	RNA-Seq of Ebolavirus infected kidney cell lines of human (HEK293T) at 6 and 24 hpi	PDGFA/B, TGFB, VEGF, CTGF GFA/B, TGFB, VEGF, CTGF, CYR6, THBS1, NOV, NRCAM, FOS, JUN, ITGB3, ITGA8, MMP2, CDH1, FN1, VCAN etc. genes were upregulated. Genes for regulation of coagulation were downregulated. IFITI were induced by Type I interferon stimulation. TFBS in AP1 were upregulated.		(Wynne et al. 2017)
Mice (A/J and C57BL/6 mice strain) /Susceptible and resistant *Staphylococcus aureus*	dual RNA-Seq of *Staphylococcus aureus* infected kidney cell at 48hpi	IL-6, IL-1a, IL-1b and TNF-a, Cxcl1, Cxcl2, Cxcl3, Saa1, Saa2, S100a8 and S100a9 genes upregulated and Inflammation-related genes, gene encoding coagulation activator tissue factor (F3) and fibrinolysis inhibitor PAI-1, Arg1 and Arg2 also highly upregulated in both. Upregulation of Hif1a was observed only in in A/J mice.	Arc operons were highly expressed during A/J infection. tcyC, lysP2, gltB, gltD, dhoM, staphopain, serine proteases, staphostatin B, aureolysin, cysteine protease, isaB, efb, hla, psma1, psma2 and psma3, RNAIII, SarR, mprF, vraX, ssaA, lytM, amiD2 and isaA), gpsB and sle1, groEL, ltaS genes highly expressed during C57BL/6 infection than A/J infection	(Thänert et al. 2017)
Entamoeba histolytica	RNA-Seq analysis of serum starved cells and serum replenishment cells		Seven motifs significantly enriched in high expression are found linked with ribosomal protein. In serum starve cells, maximum up regulated genes involved in DNA repair, BspA and Myb- mediated signalling, and lipid metabolism.	(Naiyer et al., 2019; Naiyer et al., 2019; Naiyer et al., 2021)

Host/pathogen	Method used	Findings in host	Findings in microbes	Reference
Sheep/visna maedi virus (VMV)	RNA-Seq analysis of VMV infected lung tissue of sheep	DE miRNAs like oar-miR-21, oar-miR-148a, oar-miR-379-5p, oar-let-7f and oar-miR-200a were upregulated in infected host cells whereas oar-miR-125b, oar-let-7b, oar-miR-181a, and oar-miR-30c and oar-miR-99a DE miRNAs were downregulated in infected host cells		(Bilbao-Arribas et al. 2019)

Table 2. Overview of some plant-pathogen interaction studies using transcriptomic analysis from already published data

Host/pathogen	Method used	Findings in host	Findings in microbes	Reference
Oryza sativa/ Magnaporthe oryzae	dual RNA-Seq of *M. oryzae* infected rice leaf at 24hpi	Pathogenesis-related (PR) genes involved in phytoalexin biosynthesis upregulated	MAS3(*Magnaporthe* appressoria-specific) protein-coding genes, cutinase genes, glycosyl hydrolase gene and HsbA domain containing proteins genes were upregulated	(Kawahara et al. 2012)
Soybean/Soybean mosaic virus (SMV)	RNA-Seq analysis from SMV infected soybean plants after 10 days incubation under normal light and shade	Genes involved in plant defence are upregulated under normal light but down regulated in shade. Genes involved in plant-pathogen interaction, secondary metabolism, sugar metabolism, vitamin metabolism were getting repressed in both conditions. Genes involved in signalling pathways like salicylic acid (SA), jasmonic acid (JA), and ethylene (ETH) were downregulated in shade.		(L. Zhang et al. 2019)
Fragaria × ananassa/ Xanthomonas fragariae	dual RNA-Seq of *X. fragariae* infected *F. × ananassa* leaves	Genes involved in photosystems I and II, metabolic processes, and transmembrane transports, defence response and response to biotic stimulus were upregulated at early stage. Genes involved in pathogen recognition, plant defence and disease resistance were upregulated in later time point.	At early stage of infection genes involved in host interaction, recognition, pathogenesis, biosynthetic process, translation, metabolic process, and generation of precursor metabolites were highly expressed than later time point. Putative RTX related-toxin, hemolytic and cytolytic RTX-toxins were more expressed at later stage.	(Gétaz et al. 2020)
Eucal-yptus nitens/Phytoph-thora cinnamomi	dual RNA-Seq of *P. cinnamomi* infected stem at 5 dpi (day post infection)	Genes involved in JA and ET signalling, Phenylpropanoid pathway, water deprivation and putative PR genes were upregulated. Genes involved in growth, cell wall modifications and cell wall chemistry were downregulated.	CRN family protein genes were highly expressed but genes involved in pathogenicity, virulence were downregulated.	(Meyer et al. 2016)

Host/pathogen	Method used	Findings in host	Findings in microbes	Reference
Carica papaya/ Combined infection by Papaya Meleira Virus (PMeV) and Papaya Meleira Virus 2 (PMeV2)	Transcriptomic analysis from PMeV complex infected *Carica papaya* at pre-and post-flowering	At pre-flowering stage, genes involved with stress and transport, NPR1-inhibitor, UDP-glucosyltransferases, and ethylene pathway were up-regulated while metabolism-related genes were showed downregulated. At post flowering stage several salicylic acid (SA)-activated genes, including PR1, PR2, PR5, WRKY transcription factors, ROS and callose genes were downregulated but BSMT1 and JA metabolism related genes were upregulated		(Madroñero et al. 2018)
Sugarcane Resistant (YT94/128) and susceptible (GT37)/ *Fusarium verticillioides*	Transcriptomic analysis from *F. verticillioides* infected leaves 14dpi by RNA-Seq	DEGs were mainly involved in DNA replication and metabolism, kinase, phosphotransferase, and nucleotide transferase activity. In 'YT 94/128', phenylpropanoid metabolism, oxidative phosphorylation, carbon metabolism, glutathione metabolism, proteasome, cutin, suberine, TCA cycle, amino acid metabolism and in 'GT 37,' phenylpropanoid metabolism, cutin, suberine, wax biosynthesis, linoleic acid metabolism, amino sugar and nucleotide sugar metabolism, alanine, aspartate, glutamate metabolism, pentose, glucuronate interconversions, glutathione metabolism, stilbenoid, diarylheptanoid and gingerol biosynthesis were enriched.		(Z. Wang et al. 2019)
Aegilops variabilis/ Cereal cyst nematode	RNA- Seq analysis of the root of CCN-infected *A. variabilis* at 30hpi, 3dpi, 9dpi	A total of 7,408 individual unigenes were associated with plant defence and resistance. These unigenes were classified into five GO sub-categories, three pathways, and a COG function group. These three pathways (3712 unigenes) related to plant defence, like plant-pathogen interaction pathway, phosphatidylinositol signaling system pathway and ABC transporters pathway. Also 839 unigenes were found having high homology to nematode species genes.		(Xu et al. 2012)

Table 2. (Continued)

Host/pathogen	Method used	Findings in host	Findings in microbes	Reference
Resistant (161) and Succeptible (151) Rice/ *Burkholderia glumae*	Transcriptomic analysis of leaves and stems after 48 hrs *Burkholderia glumae* infection	Up-regulated genes (456) were less than down-regulated (855) in comparison to R Vs. S. Also in S genotype, up-regulated genes (3,780) were more than R genotype (2,340) in comparison to control. Up-regulated transcripts were found in chromosomes 8 and 11 in R Vs. S comparison like disease-resistance related transcripts RPM1, NBS-LRR type and PIF-like ORF1. Genes involved in lipid metabolic process, carbohydrate binding, signal transducer, receptor activities, plasma membrane and other membrane parts were up regulated in S Vs. R manner. Genes related to cell death, nucleotide binding specifically ATP, protein binding and mitochondrion were upregulated in R Vs. S manner.		(Magbanua et al. 2014)
Resistant and susceptible wheat plants/ *Blumeria graminis f.* sp. *Tritici*	Transcriptomic analysis of Bgt infected wheat leaves at 24 hpi by gene chip to understand Pmhym-mediated resistance	164 transcripts involved in disease/defense, energy, cell structure, metabolism, secondary metabolism, protein synthesis, protein destination and storage, signal transduction, transcription, transport, and growth were induced in incompatible interaction. In compatible interaction, 198 transcripts were induced which were probably related to establishment of a biotrophic interaction with Bgt. Genes involved in salicylic acid (SA), jasmonate (JA) and ethylene (ET) signaling were differential expressed suggesting involvement in the defensive response against Bgt infection. Abscisic acid (ABA) signalling pathway genes were also upregulated suggesting inhibitory effect on SA, JA and ET signalling in susceptible reaction. Disease/defence, signal transduction and reactive oxygen species (ROS) genes were upregulated in incompatible interactions reflecting involvement in resistant response.		(J. M. Wang et al. 2012)

Host/pathogen	Method used	Findings in host	Findings in microbes	Reference
Chrysanthemum/ Cucumber Mosaic Virus (CMV), Tomato Spotted Wilt Virus (TSWV) and Potato Virus X (PVX)	A transcriptomic analysis from 5 days infected leaves of chrysanthemum by 135K microarray derived from expressed sequence tags.	33 genes were commonly found in three conditions. Among them, 22 genes were up-regulated like genes related to calcium-binding protein, glycosyltransferase family protein and an ERF TF while 11 genes were downregulated like genes related to xyloglucan-specific fungal endoglucanase inhibitor protein precursor and ribonucleoside-diphosphate reductase small chain-like protein. Six GO terms including stress response and MCM complex were commonly found in three conditions. Stresses and defence response genes were upregulated in three condition like chitin response genes (ERF TFs and a WRKY11 TF). Genes involved in DNA replication and histone modification were down regulated in three condition like MCM complex and MAD2.		(Choi et al. 2015)

3.1.3. Prediction of Biomarkers in Host Pathogen Interaction

Transcriptomic analysis helps in the prediction of biomarkers present in host pathogen interaction which helps in the diagnosis of diseases and the development of drugs. Biomarkers can be from both host and pathogen (Dix et al. 2016). *Aspergillus* cell wall component galactomannan act as a biomarker in invasive aspergillosis (Rogers et al. 2013). oar-miR-21 can be a potential biomarker during SRLV infection of sheep lung tissue (Bilbao-Arribas et al. 2019). AP1 transcription factor in human kidney cells can also be a potential biomarker during Ebola virus infection (Wynne et al. 2017). Serum miRNAs (hsa-miR-196b and hsa-miR-376c) can act as biomarkers for active TB disease in humans (H. Zhang et al. 2014) DC-STAMP and matrix metalloproteases can be potential biomarkers for leprosy in human (Dalby et al. 2020). IL-27 is a novel biomarker for bacterial infection in children (Wong et al. 2012).

Conclusion: Future Perspectives

Since the evolution of human life upon earth, both animal and plant parasitic diseases have continued to emerge and re-emerge with higher ferocity. Parasites have also developed resistance against the anti-parasitic molecules which were once useful against them. They continue to pose a serious threat to both humans and plants. In a bid to win the battle against pathogens, humankind has continuously developed technologies like transcriptomics. Transcriptomic tools have been successfully employed in host-pathogen interaction studies to decipher the insights of the changes going on during the host-pathogen interaction. More specifically, the advanced transcriptomics techniques such as Dual RNAseq and single cell RNAseq allow us to understand the changing gene expression profiles in microbial pathogens as well as in hosts during the development of the different associations. These methodologies will be a foremost tool in the management of the upsurge of microbial infections. Integration of the transcriptomics with other omics technologies like metabolomics, proteomics, etc. have been very useful in untangling genetic, metabolic, post-transcriptional, post-translational mechanisms going on during the host pathogen interaction dynamics. This integration of transcriptomics with other omics like proteomics studies improves the authenticity of the data. Thus, transcriptomics is a powerful tool for studying spatial gene expression changes enabling us with the insights of the molecular basis of host-pathogen interaction.

References

Ambriz-Aviña, Verónica, Jorge A. Contreras-Garduño, and Mario Pedraza-Reyes. 2014. "Applications of Flow Cytometry to Characterize Bacterial Physiological Responses." *BioMed Research International* 2014. https://doi.org/10.1155/2014/ 461941.

Ansari, Sekhu, Dharmendra Nath Bhatt, Chandni Sood, and Asis Datta. 2021. "Functional Characterization of the LdNAGD Gene in *Leishmania donovani*." *Microbiological Research* 251 (October): 126830. https://doi.org/10.1016/J.MICRES.2021.126830.

Bhatt, Dharmendra Nath, Sekhu Ansari, Anil Kumar, Sumit Ghosh, Alka Narula, and Asis Datta. 2020. "*Magnaporthe oryzae* MoNdt80 Is a Transcriptional Regulator of GlcNAc Catabolic Pathway Involved in Pathogenesis." *Microbiological Research* 239 (October): 126550. https://doi.org/10.1016/J.MICRES.2020.126550.

Borawski, K, Dunaj J, Pancewicz S, Król M, Czupryna P, and Moniuszko-Malinowska A. 2020. "*Coxiella burnetii* and Q Fever - a Review." *Przeglad Epidemiologiczny* 74 (1): 43–48. https://doi.org/10.32394/PE.74.05.

Casadevall, Arturo, and Liise Anne Pirofski. 2000. "Host-Pathogen Interactions: Basic Concepts of Microbial Commensalism, Colonization, Infection, and Disease." *Infection and Immunity* 68 (12): 6511–18. https://doi.org/10.1128/IAI.68.12.6511-6518.2000.

Chattopadhyay, Pratip K., Mario Roederer, and Diane L. Bolton. 2018. "A Deadly Dance: The Choreography of Host–Pathogen Interactions, as Revealed by Single-Cell Technologies." *Nature Communications* 9 (1): 1–11. https://doi.org/10.1038/s41467-018-06214-0.

Chetouhi, Chérif, Johan Panek, Ludovic Bonhomme, Hicham El Alaoui, Catherine Texier, Thierry Langin, Charissa de Bekker, et al. 2015. "Cross-Talk in Host-Parasite Associations: What Do Past and Recent Proteomics Approaches Tell Us?" *Infection, Genetics and Evolution* 33: 84–94. https://doi.org/10.1016/j.meegid. 2015.04.015.

Clemens, D. L., B. Y. Lee, and M. A. Horwitz. 1995. "Purification, Characterization, and Genetic Analysis of *Mycobacterium tuberculosis* Urease, a Potentially Critical Determinant of Host-Pathogen Interaction." *Journal of Bacteriology* 177 (19): 5644–52. https://doi.org/10.1128/jb.177.19.5644-5652.1995.

Cummings, C. A., and D. A. Relman. 2000. "Using DNA Microarrays to Study Host-Microbe Interactions." *Emerging Infectious Diseases* 6 (5): 513. https://doi.org/10. 3201/EID0605.000511.

Eberwine, James, Hermes Yeh, Kevin Miyashiro, Yanxiang Cao, Suresh Nair, Richard Finnell, Martha Zettel, and Paul Coleman. 1992. "Analysis of Gene Expression in Single Live Neurons." *Proceedings of the National Academy of Sciences of the United States of America* 89 (7): 3010–14. https://doi.org/10.1073/pnas.89.7.3010.

Eisenreich, Wolfgang, Thomas Dandekar, Jürgen Heesemann, and Werner Goebel. 2010. "Carbon Metabolism of Intracellular Bacterial Pathogens and Possible Links to Virulence." *Nature Reviews Microbiology* 8 (6): 401–12. https://doi.org/10.1038/nrmicro2351.

Fields, Chris, Mark D. Adams, Owen White, and J. Craig Venter. 1994. "How Many Genes in the Human Genome?" *Nature Genetics 1994 7:3* 7 (3): 345–46. https://doi.org/10.1038/ng0794-345.

Gengenbacher, Martin, and Stefan H. E. Kaufmann. 2012. "*Mycobacterium tuberculosis*: Success through Dormancy." *FEMS Microbiology Reviews* 36 (3): 514–32. https://doi.org/10.1111/j.1574-6976.2012.00331.x.

Glatter, Kathryn A., and Paul Finkelman. 2021. "History of the Plague: An Ancient Pandemic for the Age of COVID-19." *American Journal of Medicine* 134 (2): 176–81. https://doi.org/10.1016/j.amjmed.2020.08.019.

González-Lamothe, Rocío, Gabriel Mitchell, Mariza Gattuso, Moussa S. Diarra, François Malouin, and Kamal Bouarab. 2009. "Plant Antimicrobial Agents and Their Effects on Plant and Human Pathogens." *International Journal of Molecular Sciences 2009, Vol. 10, Pages 3400-3419* 10 (8): 3400–3419. https://doi.org/10.3390/IJMS10083400.

Goss, Erica M., Javier F. Tabima, David E. L. Cooke, Silvia Restrepo, William E. Frye, Gregory A. Forbes, Valerie J. Fieland, Martha Cardenas, and Niklaus J. Grünwald. 2014. "The Irish Potato Famine Pathogen *Phytophthora infestans* Originated in Central Mexico Rather than the Andes." *Proceedings of the National Academy of Sciences of the United States of America* 111 (24): 8791–96. https://doi.org/10.1073/pnas.1401884111.

Gross, Andre, Jonas Schoendube, Stefan Zimmermann, Maximilian Steeb, Roland Zengerle, and Peter Koltay. 2015. "Technologies for Single-Cell Isolation." *International Journal of Molecular Sciences* 16 (8): 16897–919. https://doi.org/10.3390/ijms160816897.

Haidinger, W, M P Szostak, W Beisker, and W Lubitz. 2001. "*Green Fluorescent Protein (GFP)-Dependent Separation of Bacterial Ghosts From Intact Cells by FACS.*" https://doi.org/10.1002/1097-0320.

Hu, Ping, Wenhua Zhang, Hongbo Xin, and Glenn Deng. 2016. "Single Cell Isolation and Analysis." *Frontiers in Cell and Developmental Biology* 0 (OCT): 116. https://doi.org/10.3389/FCELL.2016.00116.

Huang, Honglin, Farhan Ullah, Dao-Xiu Zhou, Ming Yi, and Yu Zhao. 2019. "Mechanisms of ROS Regulation of Plant Development and Stress Responses." *Frontiers in Plant Science* 0 (May): 800. https://doi.org/10.3389/FPLS.2019.00800.

Hwang, Byungjin, Ji Hyun Lee, and Duhee Bang. 2018a. "Single-Cell RNA Sequencing Technologies and Bioinformatics Pipelines." *Experimental and Molecular Medicine* 50 (8): 1–14. https://doi.org/10.1038/s12276-018-0071-8.

Izda, Vladislav, Matlock A. Jeffries, and Amr H. Sawalha. 2021. "COVID-19: A Review of Therapeutic Strategies and Vaccine Candidates." *Clinical Immunology* 222 (January): 108634. https://doi.org/10.1016/J.CLIM.2020.108634.

Kaper, James B., James P. Nataro, and Harry L. T. Mobley. 2004. "Pathogenic *Escherichia coli*." *Nature Reviews Microbiology 2004 2:2* 2 (2): 123–40. https://doi.org/10.1038/nrmicro818.

Kresge, Nicole, Robert D. Simoni, and Robert L. Hill. 2006. "The Purification and Sequencing of Alanine Transfer Ribonucleic Acid: The Work of Robert W. Holley." *Journal of Biological Chemistry* 281 (7): e7–9. https://doi.org/10.1016/S0021-9258(19)48443-3.

LaMonte, Gregory M., Pamela Orjuela-Sanchez, Jaeson Calla, Lawrence T. Wang, Shangzhong Li, Justine Swann, Annie N. Cowell, et al. 2019. "Dual RNA-Seq Identifies Human Mucosal Immunity Protein Mucin-13 as a Hallmark of Plasmodium Exoerythrocytic Infection." *Nature Communications 2019 10:1* 10 (1): 1–13. https://doi.org/10.1038/s41467-019-08349-0.

Leroy, Quentin, and Didier Raoult. 2010. "Review of Microarray Studies for Host–Intracellular Pathogen Interactions." *Journal of Microbiological Methods* 81 (2): 81–95. https://doi.org/10.1016/J.MIMET.2010.02.008.

Lowe, Rohan, Neil Shirley, Mark Bleackley, Stephen Dolan, and Thomas Shafee. 2017. "Transcriptomics Technologies." *PLOS Computational Biology* 13 (5): e1005457. https://doi.org/10.1371/JOURNAL.PCBI.1005457.

Ma, Haijie, Bin Zhang, Yunpeng Gai, Xuepeng Sun, Kuang Ren Chung, and Hongye Li. 2019. "Cell-Wall-Degrading Enzymes Required for Virulence in the Host Selective Toxin-Producing *Necrotroph alternaria alternata* of Citrus." *Frontiers in Microbiology* 10 (November): 1–15. https://doi.org/10.3389/fmicb.2019.02514.

Mamanova, Lira, Zhichao Miao, Ayesha Jinat, Peter Ellis, Lesley Shirley, and Sarah A. Teichmann. 2021. "High-Throughput Full-Length Single-Cell RNA-Seq Automation." *Nature Protocols 2021 16:6* 16 (6): 2886–2915. https://doi.org/10.1038/s41596-021-00523-3.

Mansueto, Pasquale, Giustina Vitale, Antonio Cascio, Aurelio Seidita, Ilenia Pepe, Antonio Carroccio, Salvatore Di Rosa, Giovam Battista Rini, Enrico Cillari, and David H Walker. 2012. "New Insight into Immunity and Immunopathology of Rickettsial Diseases." *Clinical and Developmental Immunology* 2012. https://doi.org/10.1155/2012/967852.

Martin-Gayo, Enrique, and Xu G. Yu. 2019. "Role of Dendritic Cells in Natural Immune Control of HIV-1 Infection." *Frontiers in Immunology* 10 (JUN): 1–9. https://doi.org/10.3389/fimmu.2019.01306.

McCall, Andrew D., Ruvini U. Pathirana, Aditi Prabhakar, Paul J. Cullen, and Mira Edgerton. 2019. "*Candida albicans* Biofilm Development Is Governed by Cooperative Attachment and Adhesion Maintenance Proteins." *Npj Biofilms and Microbiomes 2019 5:1* 5 (1): 1–12. https://doi.org/10.1038/s41522-019-0094-5.

McKinnon, Katherine M. 2018. "Flow Cytometry: An Overview." *Current Protocols in Immunology* 120 (1): 5.1.1-5.1.11. https://doi.org/10.1002/CPIM.40.

Mogensen, Trine H. 2009. "Pathogen Recognition and Inflammatory Signaling in Innate Immune Defenses." *Clinical Microbiology Reviews* 22 (2): 240–73. https://doi.org/10.1128/CMR.00046-08.

Moreno-Campuzano, Samadhi, Sarath Chandra Janga, and Ernesto Pérez-Rueda. 2006. "Identification and Analysis of DNA-Binding Transcription Factors in *Bacillus subtilis* and Other Firmicutes- a Genomic Approach." *BMC Genomics 2006 7:1* 7 (1): 1–10. https://doi.org/10.1186/1471-2164-7-147.

Naiyer, S., Bhattacharya, A., Bhattacharya, S., 2019. Advances in *Entamoeba histolytica* biology through transcriptomic analysis. *Front. Microbiol.* 10, 1921. https://doi.org/10.3389/FMICB.2019.01921/BIBTEX.

Naiyer, S., Kaur, D., Ahamad, J., Singh, S. S., Singh, Y. P., Thakur, V., Bhattacharya, A., Bhattacharya, S., 2019. Transcriptomic analysis reveals novel downstream regulatory motifs and highly transcribed virulence factor genes of *Entamoeba histolytica.* BMC Genomics 20, 1–18. https://doi.org/10.1186/S12864-019-5570-Z/TABLES/6.

Naiyer, S., Singh, S. S., Kaur, D., Singh, Y. P., Mukherjee, A., Bhattacharya, A., Bhattacharya, S., 2021. Transcriptomic analysis of ribosome biogenesis and pre-rRNA processing during growth stress in *Entamoeba histolytica.* bioRxiv 2021.08.01.454488. https://doi.org/10.1101/2021.08.01.454488.

Nuss, Aaron M., Michael Beckstette, Maria Pimenova, Carina Schmühl, Wiebke Opitz, Fabio Pisano, Ann Kathrin Heroven, and Petra Dersch. 2017. "Tissue Dual RNA-Seq Allows Fast Discovery of Infection-Specific Functions and Riboregulators Shaping Host–Pathogen Transcriptomes." *Proceedings of the National Academy of Sciences* 114 (5): E791–800. https://doi.org/10.1073/PNAS.1613405114.

Quainoo, Scott, Jordy P. M. Coolen, Sacha A. F. T. van Hijum, Martijn A. Huynen, Willem J. G. Melchers, Willem van Schaik, and Heiman F. L. Wertheim. 2017. "Whole-Genome Sequencing of Bacterial Pathogens: The Future of Nosocomial Outbreak Analysis." *Clinical Microbiology Reviews* 30 (4): 1015–63. https://doi.org/10.1128/CMR.00016-17.

Radwanska, Magdalena, Nick Vereecke, Violette Deleeuw, Joar Pinto, and Stefan Magez. 2018. "Salivarian Trypanosomosis: A Review of Parasites Involved, Their Global Distribution and Their Interaction With the Innate and Adaptive Mammalian Host Immune System." *Frontiers in Immunology* 0 (OCT): 2253. https://doi.org/10.3389/FIMMU.2018.02253.

Saizieu, Antoine de, Ulrich Certa, Janet Warrington, Christopher Gray, Wolfgang Keck, and Jan Mous. 1998. "Bacterial Transcript Imaging by Hybridization of Total RNA to Oligonucleotide Arrays." *Nature Biotechnology 1998 16:1* 16 (1): 45–48. https://doi.org/10.1038/nbt0198-45.

Scarpa, Raffaele, Francesco Caso, Luisa Costa, Saverio Passavanti, Maria Grazia Vitale, Claudia Trojaniello, Antonio Del Puente, and Paolo A. Ascierto. 2020. "May the Analysis of 1918 Influenza Pandemic Give Hints to Imagine the Possible Magnitude of Corona Virus Disease-2019 (COVID-19)?" *Journal of Translational Medicine* 18 (1): 1–12. https://doi.org/10.1186/s12967-020-02673-6.

Sen, R., L. Nayak, and R. K. De. 2016. "A Review on Host–Pathogen Interactions: Classification and Prediction." *European Journal of Clinical Microbiology and Infectious Diseases* 35 (10): 1581–99. https://doi.org/10.1007/s10096-016-2716-7.

Sena, Johnny A., Giulia Galotto, Nico P. Devitt, Melanie C. Connick, Jennifer L. Jacobi, Pooja E. Umale, Luis Vidali, and Callum J. Bell. 2018. "Unique Molecular Identifiers Reveal a Novel Sequencing Artefact with Implications for RNA-Seq Based Gene Expression Analysis." *Scientific Reports 2018 8:1* 8 (1): 1–13. https://doi.org/10.1038/s41598-018-31064-7.

Tang, Fuchou, Catalin Barbacioru, Yangzhou Wang, Ellen Nordman, Clarence Lee, Nanlan Xu, Xiaohui Wang, et al. 2009. "MRNA-Seq Whole-Transcriptome Analysis of a Single Cell." *Nature Methods* 6 (5): 377–82. https://doi.org/10.1038/ nmeth.1315.

Valihrach, Lukas, Peter Androvic, and Mikael Kubista. 2018. "Platforms for Single-Cell Collection and Analysis." *International Journal of Molecular Sciences* 19 (3): 22–24. https://doi.org/10.3390/ijms19030807.

Weitzman, Matthew D, and Amélie Fradet-Turcotte. 2018. "Annual Review of Virology Virus DNA Replication and the Host DNA Damage Response." *Annu. Rev. Virol* 5: 141–64. https://doi.org/10.1146/annurev-virology.

Westermann, Alexander J., Lars Barquist, and Jörg Vogel. 2017. "Resolving Host–Pathogen Interactions by Dual RNA-Seq." *PLoS Pathogens* 13 (2): 1–19. https://doi.org/10.1371/journal.ppat.1006033.

Westermann, Alexander J., Stanislaw A. Gorski, and Jörg Vogel. 2012. "Dual RNA-Seq of Pathogen and Host." *Nature Reviews Microbiology* 10 (9): 618–30. https://doi.org/10.1038/nrmicro2852.

Willyard, Cassandra. 2018. "New Human Gene Tally Reignites Debate." *Nature 2021 558:7710*, June. https://www.nature.com/articles/d41586-018-05462-w.

Wong, William Chi Wai, Stephanie Tsz Hei Lau, Edmond Pui Hang Choi, Joseph D Tucker, Christopher K Fairley, and John M Saunders. 2019. "A Systematic Literature Review of Reviews on the Effectiveness of Chlamydia Testing." *Epidemiologic Reviews* 41 (1): 168–75. https://doi.org/10.1093/EPIREV/MXZ007.

Zuzarte-Luís, Vanessa, and Maria M. Mota. 2018. "Parasite Sensing of Host Nutrients and Environmental Cues." *Cell Host & Microbe* 23 (6): 749–58. https://doi.org/10.1016/J.CHOM.2018.05.018.

Chapter 7

Transcriptomics in Microbial Bioremediation

Bisma Rashid*, Saima Shafi, Malik Asif Aziz, Zaffar Mehdi Dar, Khalid Hussain Bhat and Mushtaq Ahmad Malik
Division of Basic Sciences and Humanities, FoA, Wadura, Sopore, SKUAST-Kashmir, India

Abstract

Effective restoration in environment is an important aspect in present day scenario and for that purpose the one among the best method considered is the bioremediation mediated by microbes, but inefficient knowledge regarding the same is a halt in the on field application of such techniques. Transcriptomics, proteomics, and interactomics, which are all relatively new approaches, hold a lot of potential as tools for addressing long-standing issues about the molecular mechanisms that drive mineralization routes. Transcript architecture and expression during mineralization have been examined utilising high-throughput transcriptomic approaches with microarrays. As far as the concern is related to the studying of mRNA at expression level, DNA microarray is considered to be one of the most sophisticated among others. Two-dimensional polyacrylamide gel electrophoresis in proteomics is a good method used in proteomics to determine the proteins (2-DE). For the construction of functional proteomics, the most recent breakthrough is the mass spectrometry and protein microarray. By far bioremediation mediated through use of differential expression can help in determining the whole profile of protein of microbes, essentially important for the purpose of bioremediation mediated by microbes.

---

*Correspondence to be addressed: bismarashidbhat@gmail.com.

In: Transcriptomics and their Importance
Editors: Sarah Naiyer and. Khalid Hussain Bhat
ISBN: 979-8-88697-044-9

Keywords: transcriptomics, bioremediation, proteomics, environmental cleanup, pollutants

1. Introduction

Due to industrialization and population consequences growth, pollution levels are rising at an alarming rate. The usage and manufacturing of synthetic substances regarding arduous advances followed by formation of imperishable contaminants viz: hydrocarbons, heavy metals, xenobiotics other pollutants as a result of industrialisation (Labie 2007). These harmful pollutants often remain embedded in the habitat since a much enhanced prolonged time and constitute a major hazard to living things. As public awareness has grown, a slew of new ways based on cutting-edge scientific technologies have emerged to assess and address this difficult worldwide problem. The extensively used method for cleansing a dirty habitat in an environmentally benign as well as long-term way is biodegradation a thing? (Paul 2005, Raghunandan 2018). Biodegradation is a fascinating method for removing toxins from the environment. As most of the important microorganisms viz: yeast, bacteria, and fungi being long since proved superior microbes for pollutant abrupt withdrawal (Zhong and Zhou, 2002, Luciene 2015, Abou Seeda 2017). This ensures low-cost, straightforward, and environmentally amiable tidy up approach (Lovely 2003). As most of the microbes have the ability to adapt to harsh environmental circumstances and are nutritionally versatile. They also possess a variety of intracellular and extracellular enzymes that are using compound defilement and change them to carbon as well as energy (Nojiri and Tsuda 2005, for example). Thakur and colleagues (Thakur 2019). All go through a quick heredity metamorphosis that allows these to gain improved anabolic pathways as xenobiotic degradation (Poirier 2013, Igiri 2018). Off-site conservation or on-situ conservation microbial biodegradation techniques are available. Off site techniques involve transferring polluted substances from one location to alternative as therapy, whereas various in situ treatments treat various polluted substances on the spot. Many academics believe that ex-situ remediation is not a viable option for cleanup due of its limitations. It may or may not be profitable at specific locations, and it's likely that the microbes that helped clean up contaminants under these conditions are no longer useful. Under in-vivo conditions, in-vitro circumstances decline to effectively eradicate them (Head 2003, Barupal 2019). In order to have a proper understanding of research which is going to have a vast impact in the proliferation of microbes in polluted habitat (Lovely, 2003). As a result, substantial study on s microbes harmony is required as to examine the bioremediating potential of microbes found near contaminated environments, as well as alterations in their form and anabolic pathways. Since microbial biodegradation solutions rely upon a wide group of microorganisms which are native to the contaminated area have an enormous anabolic power. The study of microorganisms reveals some information on microbial anabolics and various degeneration processes. Because most of bulk of microorganisms residing the habitat are not able to culture under outside the living conditions, the approach for accessing the microbial world remains baffling (Handelsman, 2004). Only a small percentage of microbes regarding various environmental perspectives are assiduously cultivable without difficulty, making them unavailable for basic study (Dickson 2014, Bursle and Robson 2016, Awasthi 2020). New improvements in genetic technologies have given researchers a competitive advantage in evaluating uncultivable species being observed from various essential habitats (Handelsman 2004, Gilbert and Dupont 2011). Genomic, proteomic, transcriptomic, metabolomic, fluxomic, and other molecular methods have provided deeper insight into the microbiome that inhabit a certain habitat (De Sousa 2018, Gupta 2020, Rodriguez 2020). Researchers have been able to overcome the challenges connected with unculturable bacteria thanks to the introduction of new-generation molecular sequencing genetic techniques (Maphosa 2010, Malla 2018). Since various advancements been aided in collecting of extensive biotic data on microbes, their biological systems, various organic molecules such as various protein enzymes involved in bioremediation processes (Zhang 2010, Misra 2018, Pandey 2019). The solution for cleaning up polluted environments is bioremediation being a long-term process. However, as there are still various issues with the technique that must be corrected. The breakdown of contaminants utilising organisms ranging from various microbes including bacteria to plants, or their derivatives, is known as bioremediation. Regarding the overall cleanup of all the polluted places across the United States is found to be $1.7 trillion (Kuiper 2004), since the biggest advantage of bioremediation over the traditional procedures like incineration is its lower cost. Understanding the synergy with the environment, and the accompanying elements regarding the indigenous microbiome that aid in the expression of their genotype is the most important stage (Chakraborty 2012, Callaghan 2013). To make bioremediation being versatile technique for environmental monitoring and to uncover the routes and chemistry that remain unknown there is a requirement of multidisciplinary approach. For biodegradation, large number of artificial compounds proved to have absence of several enzymes (since organic compounds about 10 million explained, and majority of these being not ascertained. (Wackett, L.P. 2001)), and however, there are many examples in which several excellent enzymes breakdown to change the pollutants in the habitat (Yee, D.C. 1998). Furthermore, procedure that transfers pollutants from one phase to another, rather than a remediation process, bioremediation is frequently considered to be such kind of long-term procedure (allowing exact alteration of the contaminants to its specified constituents including water and carbon dioxide) (Kuiper 2004).

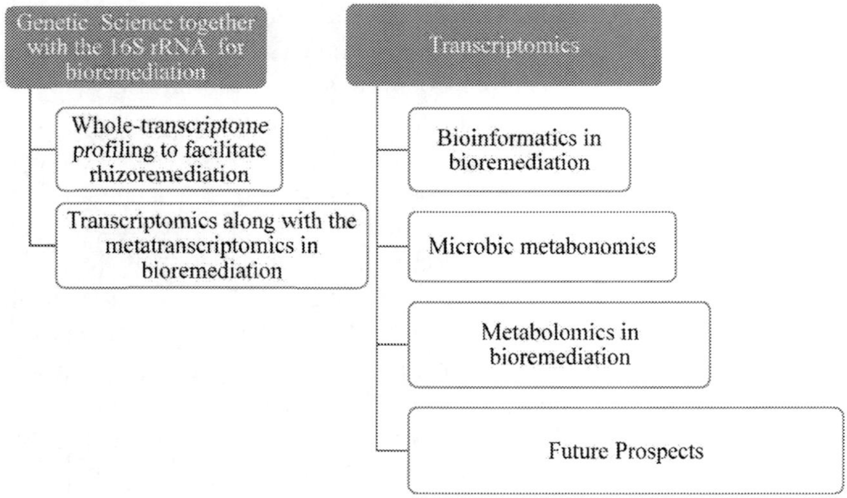

Figure 1. Graphical Representation of the various contents.

2. Genetic Science Together with the 16S rRNA for Bioremediation

Genomic technology has aided in the long-term remediation of polluted surroundings. The multitude of bacteria and the sense of community processes actively engaging in bioremediation has been enhanced by cultivation-independent studies of microorganisms from polluted areas (Kapley 2007, Desai 2010). The evolution of microbiome utilising contemporary genetic information is being observed by the process of identification of unique microbes those were not observable by accomplishment techniques. Malik 2008, Rani 2008, Tretyn and Golebiewski (2020). The evolution of the 16S being regarded the greatest quality level process for defining genetic similarities among organisms in microbiome which is considered to be effectively varied and conserved in all microbes (Lovely 1991, Lovely, 2003). The expansion and arrangement of the complementary portions of the 16S rRNA gene sequence analysis can be done in case of complete analysis of microbiome. It's a very efficient and cost-effective technique that can be accessed through a variety of bioinformatics tools, and it's becoming an excellent procedure for studying complex microbiome (Han 2020). Since it is proved to a technique that can be used to identify bacteria that are new, unculturable, or phenotypically unidentified (Clarridge, 2004). To establish the phylogenetic order of bacteria involved in bioremediation processes, the various gene sequences being used which have been collected from different polluted environments (Lovely, 2003, Rogers and McClure 2003). In order to study the research in the genetics of PAH degradation, 16S rRNA sequences being used to explain the constituents of the microbiome as well as the diversity of genes which are of deoxygenase nature in soil embedded in a coal tar mixture in order to facilitate the research study (Kumar and Khanna 2010, Viant and Sommer 2013, Sakshi Haritash 2020). In Shanghai, Kou (2018) to study the abundance and diversity of the microbial community in soil that had been the contamination with hard core rock such as copper, zinc as well as lead employing various gene amplicon arranging analysis have been employed in order to observe the diversity and abundance of microbiome in soil. From a site with a lot of fragile plastic waste, various Pseudomonas species been capable of degrading polyurethane were identified membrane fatty acid profiles and 16S rRNA molecular sequencing (Cardenas Espinosa 2020).

3. Transcriptomics

The transcriptome defines a division of an organism's genes that are transcribed. The genome, proteome, and cellular phenotype form a powerful network. It's also known as gene expression profiling since it explains how genes are up- or down-regulated in different conditions in microbiome. The various processes viz: the process which gives a straight view regarding various genetic manifestations is mRNAs analysis, such as I transcript, (ii)assessable evaluation of genotype impact on isoforms of protein, and (iii) another splicing evaluation to predict isoforms of proteins. Gene expression can be used in case of allele-specific expression or expression evaluated via trait loci. Microarray and sequencing techniques are used to study the transcriptome (Maroli 2018). Using next-generation sequencing, RNA-sequencing is used to observe the amount of RNA in the respective sample whereas, microarrays being used to analyse gene expression. Pre-designed probes are used in the RNA-microarray technology, making it a more cost-effective, potent, and improved method for protein investigating process. As RNA sequencing process offers a wide range of RNA types to be sequenced at a high level of analysis and essence research (Shendure 2008, Nagalakshmi 2010). Since, genetic expressions that are modulated by a changing environment, the evolution of DNA microarray based on whole-genome technology has allowed the researchers to study the entire process (Gao 2004). The process of evaluating the various changes that took place in anabolic and catabolic gene expressions, investigate physiology of microbiome of various habitats, and discover new bacterial species, among other things (Dennis 2003, Greene and Voordouw 2003). In case of exploring and consolidating the molecular expression (RNA data) with molecular variations (DNA data) one of the most acceptable technique is the aligning of expression quantitative trait loci(eQTL). The analysis of eQTL deals with the identification of genetic variations that control mRNA expression levels. The core of molecular networks associated with a specific feature tends to allow a link between molecular variations and gene expression thereby allowing researchers to study (Yao 2015). To investigate gene expression regulation, the mRNA analysis and microRNA levels can be employed. MicroRNAs (mRNAs) are noncoding RNA molecules that regulate mRNA transcription. Since as a result, for a wide range of molecular expression dynamics, it's been quite obvious not only just the miRNA levels but along with control by mRNAs are crucial (Pritchard 2012). In order to discover various changes in the manuscript outline, especially to discover miRNA which occurred to be regulated, highlighting the likely genetic pathways underlying a certain condition, identification of miRNA transcripts which are regulated, mRNA profiling as well as miRNA expression can be employed. This method has been effectively used in case of evaluation of molecular expression in certain species of digestive tract cancers (Pantaleo 2016), and in case of diabetes patients' wound-healing impaired fibroblasts to assess gene expression modification method is used (Liang 2016). Different transcriptomics being identified which are generally highly-upgraded involving phenol breakdown processes as well as powerful transporters, there is a need to enhance the tolerance of phenol and utilisation by accumulation of lipid by Rhodococcus opacus PD630 (Yoneda 2016). Using manuscript examination of sewage treatment microbiomes being a well translated process and as well as an efficient process which helps in determining the beneficial role of nitrogen fixing life form regarding crude oil breakdown (Sato 2019). Using the significance of differentially expressed genes associated in crude oil degradation found significant in the process regarding identification with respect to crude oil degrading strains of Pseudomonas aeruginosa (Das 2020).

3.1. Whole-Transcriptome Profiling to Facilitate Rhizoremediation

As comparison to procedures like proteomics, the advantage of being able to determine the amount of transcripts from the entire genome involves the identification of majority of proteins which may include few examples viz: proteins which are membrane-bound as well as troublesome and approximately 2/3 of E.coli proteins being not discovered by non-gel proteomic procedures, also using DNA microarrays based on profiling of whole-transcriptome (Han, M-J. and Lee, S.Y. 2006). In order to identify the differences in the process of protein production, the profiling of transcriptome oftenly predicts differences in transcriptions may be used, but it's not always as such but however, in case of prokaryotes as due to regulation process which usually occurs at the transcription level. To know the better understanding of the anabolism relationship various researchers started with respect to utilise genetic makeup. The research has been done as the first whole-transcriptome of bacteria as well as in plants along with wide variety of numerous virulence genes were identified on Erwinia chrysanthemi African violet leaves (Okinaka, Y. 2002). The evolution of seven novel bacterial virulence genes as well as the first whole-transcriptome based research being done in case of bacteria found in the rhizosphere as well as in case of poplar tree roots along with the pathogenic Pseudomonas aeruginosa were found (Attila, C. 2008). The study of symbiotic synergy that occurs around the root zone of plants along with the organisms which are beneficial for bioremediation, as well as whole-transcriptome based transcriptome profiling being utilised viz: around 90 rhizosphere highly regulated genes coding for P.putida being developing on maize roots were observed (Matilla, M.A. 2007). Bacteria have also been chosen for their ability to perform well in rhizoremediation, such as facilitating the breakdown of the pesticide lindane (g hexachlorocyclohexane) and its related chemicals and d-hexachlorocyclohexane. In order to utilise a bifold enhancement procedure Ramos and associates isolated four Sphingomonas strains which can digest lindane and which flourish around the rhizosphere of corn (Boltner, D. 2008). This method shows potential for associating bacteria with elevated deterioration capacity for plant hosts which are compatible.

Table 1. Genomics Revolution in Bioremediation

Microorganism	Bioremediation	Website
Dehalococcoides ethanogenes	Decreasing deflocculation regarding chlorinated solvents to ethylene. It's been observed that similarly associated series which are enhanced inside underground habitats which are being degraded is the consequence of protein synthesis of D. ethanogenes as well as molecular series	http://www.tigr.org
Geobacter sulfurreducens *Geobacter metallireducens*	Reductive precipitation of uranium as well as anaerobic oxidation of aromatic hydrocarbons. Since, in case of *Geobacter* species the 16S rRNA molecular series are thought to be similarly associated in paramount being observed during an anaerobic inert biodegradation of uranium and aerenes.	http://www.jgi.doe.gov http://www.tigr.org
Rhodopseudo-monas palustris	The leading living form responsible to eradicate the progression of aerenes as well as their anabolism and regulation.	http://www.jgi.doe.gov
Dechloromonas aromatica	Typical of omnipresent group of perchlorate-reducing microbes as well as being competent of benzene combined resulted in the nitrate reduction for the anaerobic oxidation.	http://www.jgi.doe.gov
Pseudomonas putida	Anabolically significant microbes competent of decomposing a broad variety of biological pollutants aerobically. Efficient living form required for molecular engineering having biodegradation potentialities.	http://www.tigr.org
Rhodopseudo-monas palustris	The leading living form for eradicating progressions of anaerobic anabolism of arenes as well as their regulation.	http://www.jgi.doe.gov
Desulfovibrio vulgaris	Exhibit to diminish bring about in chromium as well as uranium. Since, the confirmed part in polluted habitats perhaps so far to be described.	http://www.tigr.org
Deinococcus radiodurans	Found to be very contrary to emission perhaps considered molecularly structured for biodegradation of contaminated habitats.	http://www.tigr.org

Table 2. Absolute genetic data series of few microbes notable for biodegradation

Title regarding the living form	Biodegradation importance	Genome size	References
Aceinatober sp.	Capacity to alter more dissimilar arenes to usual medium	3.6Mb	Barbe et al. (2004)
Alcanivorax borkumensis SK2	Capacity of bioremediating organic compounds	3.1Mb	Schneiker et al. (2006)
Arthrobacter aurescens TC1	Capability to breakdown glyphosate, methyl ter-butyl ether, nicotine, 4-nitrophenol, dimethlysilanediol as well as dichlorophenoxyacetate	4.6Mb	Mongodin et al. (2006)
Geobacillus thermodenitrificans NG80-2	Carboniferous bioremediating archaea	3.6Mb	Feng et al. (2007)
Desulfovibrio vulgaris Hildenberough	Degradation of chromium and uranium	3.57Mb	Heidelberg et al. (2004)
Rhodopseudomonas palustris	Capacity to breakdown chemical compounds in anoxic contaminated sites	5.5Mb	Larimer et al. (2004)
Geobacter sulfurreducens PCA	Bioremediation of chemical compounds regarding anoxic contaminated sites	3.8Mb	Methe et al. (2003)
Pseudomonas mendocena ymp	Biodegradation of organochlorine compounds	5.07Mb	Unpublished (GenBank:CP000680)
Ralstonia eutropha JMP134	Bioremediation of chlorinated compounds	3.8Mb	Louie et al. (2002)
Sphingomonas wittichii RW1	Biodegradation of di-benzo-p-dioxin and dibenzofuran	5.9Mb	Unpublished (GenBank:CP000699)
Shewanella sediminis HAWEB3	Biodegradation of hexahydro-1,3.5-trinitro-1,3,5-triazine(RDX)	5.5Mb	Unpublished (GenBank:CP000821)

Table 3. Relative survey about following procreation arrangement considering microbiota modules

S.No:	Omics technologies	Principle	Advantages	Disadvantages	Applications
1.	Pyrosequencing	The identification of pyrophosphate discharged after nucleotide amalgamation in the newly synthesized DNA strand being established by sequencing synthesis	Rapid and definite way being along actual time to be exile accurate as required in arranging small particles of DNA, nucleotide dispensation, inexpensive as well as simply accessible	The development of prolonged identical polymers giving rise to arranging inaccuracy, trouble in deciding quantity of integrated nucleotides in homopolymeric areas	Discovery about microbes, whole genome sequencing
2.	Illumina	Sequence-by-synthesis method	Cells generally discretional, quick reference, give rise to comparatively brief perusal along extent all of 300bp	Aberrant incorporation of incorrect dNTPs by polymerases, short, slow, high cost of reagents, short reads	Molecular genetics scrutinize and discover isotypes, new manuscript, chimeric genes, whole exome arranging
3.	ABI SOLID system	Sequencing technology based on ligation of DNA fragments	High efficiency every base is studied over again	Generally brief reads	Entire genetic set arranging, rapid and cost effective arranging, manuscript exploration, above the genome survey
4.	Ion Torrent	Sequencing established through the recognition of hydrogen ions released during polymerization of DNA	Obtains reads all of 400 bp long, dependable and economical implement, elevated efficiency and brief time	More elevated inaccuracy assess regarding particular regions	In order to obtain microbiome inside complex habitats

S.No:	Omics technologies	Principle	Advantages	Disadvantages	Applications
5.	Heliscope	Single-molecule sequencing platform using a highly sensitive fluorescence detection system	More quantity of individual particles peruse, in order to lower down elevated inaccuracy assess	Brief perusal succession, the arranging procedure degrade out of many partiality due to insufficient duplicate expansion as well as DNA addition desolation	Resequencing, manuscript enumeration
6.	Pacific biosciences SMRT sequencing	Sequencing through synthesis method and real time detection of integrated fluorescently labelled nucleotides	Extended peruse, elevated precision, same reportage, individual particle settlement	Extending peruse generate staging in replicate zones	Entire genetic set arranging, rapid and cost effective, arranging of RNA, phenotypic discoveries, obtain composite inhabitants
7.	Oxford nanopore sequencing	Determination of material differences which occurs regarding DNA succession resulted in alteration directs nanometre sort opening down the effect of charged province	Inexpensive, rapid as well as efficient arranging of DNA, lengthy peruse, goods settlement	Partiality in inaccuracy, elevated value peruse	Scrutiny of proteins, RNA as well as DNA

4. Bioinformatics in Bioremediation

Since based on various transcriptome techniques which have been found significant, they are not able to demonstrate the differences that took place during transcription translate various differences in protein levels, which are required for better bioremediation. Being the greatest approaches regarding evaluating differences with respect to anabolism is measuring the whole differences in various levels of protein which are quite necessary for enhanced bioremediation the major pathways that are discovered evaluating the differences in anabolism is bioinformatics or in another way of measuring the total differences occurring in levels of proteins. In order to evaluate the effect of anabolic scheme on biodegradation is bioinformatics TCE analysis which is considered to be one of the first attempts. Since eight new proteins were observed in TOM-containing cells, along with 12 proteins which are not detected in the cells which were present in the host strain as well as two-dimensional electrophoresis (2-DE) utilized that involved in the conversion of TCE into a reactive TEC epoxide (Pferdeort, V.A. et al. 2003). TCE exposure resulted in the creation of only one new protein and the deletion of ten others in TOM-containing cells. As a consequence of anabolic engineering which plays a significant role as well as provides major influence on the cells, as described by a proteomic analysis.

4.1. Transcriptomics along with the Meta-transcriptomics in Bioremediation

By examining mRNA transcriptional profiles, it provides the ability to obtain various processes of habitat of microbial populations, various transcriptome or metatranscriptomics technologies being used (McGrath 2008). Various stages involved in transcriptomic analyses: (1) total mRNA extraction and enrichment, (2) Synthesis of complementary DNA, (3) Either entire complementary DNA transcriptome or based on microarray hybridization of complementary DNA. Regarding the transcriptomics analysis, it's been observed that in Polaromonas sp., the assimilation took place on cis-dichloroethene (cDNA) Jennings 2009). To determine the genes increased by cDCE in the JS666 strain DNA microarrays were used. During the development of cDCE in this investigation, the molecular expression based on genome arrays for Polaromonas sp. strain JS666 strain being developed, hybridised as well as complementary DNA extracted. The genes increased by cDCE including various ABC transporters, proteins which are antioxidant in nature as well as solute/sodium symporters were observed. Furthermore, a key degradation route involving carbon-chloride cleavage was proposed. Polaromonas sp. JS666 strain biodegraded cDCE via a major degradation mechanism including monooxygenase-catalyzed epoxidation and a pathway involving monooxygenase-catalyzed epoxidation which is comparatively minor process. The similar kind of work was studied by Holmes (2009) regarding transcriptome of the Geobacter uraniireducens strain being developing in uranium polluted subterranean deposits. The whole-genome microarray procedure which is being compared to sediment-grown being observed on G. uraniireducens the cells of which develop in particular growth medium found around 1084 genes enhanced transcript encompassing the growth found in uranium-polluted sediments. Around thirty-four c-type genes which are of cytochrome nature are found to be involved in reduction of U (VI) as well as Fe(III) found to be increased in G. uraniireducens sediment cultured cells. Popularly known as the "metatranscriptome," the researchers now found the molecular expression profile of the whole microbiome. The headon elicitation of messengerRNA from various archaeal, bacterial, and eukaryotic microbial cells are the consequences of the recent improvements. The use of metatranscriptomes in conjunction with pyrosequencing or the fabrication of complementary DNA based on microarrays allows handy equipment for monitoring the transcriptional processes of whole microbial communities (Urich 2008).

5. Microbic Metabonomics

Besides genomes, proteomics, transcriptomics, researchers are now turning their attention to the global investigation of a microbial cell's whole repertoire of cellular metabolites, a technique namely "metabolomics."The microbic cell in response to an environmental change or stress derives various primary and secondary metabolites which are of low molecular weight. The aim is to quantify the significant part regarding those substances in microbic cells involving various procedures including systematic techniques, separation as well as metabonomics. The inclusion in the microbial metabolomics toolkit involves various methods viz: metabolite profiling, anabolic fingerprinting. Since, several analytical technologies for quantification as well as identification of a wide range of cellular metabolites (Mapelli et al., 2009) includes: direct injection mass spectrometry(DIMS), nuclear magnetic resonance (NMR), high performance liquid chromatography(HPCL), gas chromatography(GC), fourier transform-infrared(FT-IR) spectroscopy, capillary electrophoresis(CE), profiling and target analysis as well as mass based spectrometry. Various recent research has used microbiome metabonomic evaluation to investigate the bioremediation of human-induced contaminants. Keum (2009) recently published research on relative metabonome procedure during phenanthrene breakdown of *Sinorhizobium sp.* C4. Since, intracellular metabolomes phenanthrene breakdown being related to that of natural resources. However, during the metabolite profiles investigation, including fatty acids, polar metabolites as well as polyhydroxyalkanoates in which nontargeted metabolome examinations being used. Around 207 peaks comprising 60% found in the study of Sinorhizobium sp. Since, C4 antagonistic breakdown product fractions were identified and compared by these researchers. The metabolic profiles collected in this work show that Sinorhizobium sp. accumulated many intermediates viz: branched-chain amino acids, trehalose, glycolysis intermediates as well as tricarboxylic acid cycle. During the metabolism of phenanthrene, C4 decreased in comparison to the other 70% of detected metabolites. The importance of metabolomics observations in bioremediation research is demonstrated in studies like these. The fact that anabolism is a controlled system at the genomic, transcriptional, and posttranslational levels are now well accepted. As a result, a relatively new concept known as "fluxomics" has been presented, which focuses flux examination based on real-time of biological particles/substances inside the cell above time (Wiechert 2007). Tang (2009) however, did an examination of fluxomics on species of Shewanella which is found to be hazardous metals and radionuclide bioremediation containing co-metabolite progression.

6. Metabolomics in Bioremediation

As described in the preceding sections, to explain or correspond genetic and genomics data, many in pipelines, web resources, silico softwares as well as algorithms have been found. Since, there are few informatics resources dedicated solely and wholly confined to bioremediation. The first and most widely used web-based resource for microbial bioremediation was observed at the University of Minnesota Biocatalysis/Biodegradation Database(UMBBD) with 187 pathways, 1287 reactions, 1195 compounds,

833 enzymes, 491 microorganism entries, and 259 biotransformation rules. However, UMBBD is readily accessible web based at (http://umbbd.msi.umn.edu/) (Ellis 2006). In order to keep track of heterogeneous information metarouter is one more method which offers question adjustment and recondition based on biodegradation as well as bioremediation. Since, system's heart is conceptual information that stores data on synthetic compounds, processes, catalyst as well as life form in a unified structure. It enables the discovery of routes connecting two chemicals. However, depiction of alternative progression perhaps limited based on variety of factors such as pathway's extent, the essential catalysts commenced perhaps in a given organism(s), intervening molecules with a scope of advantages for a specified attribute and so at length. A web interface can be used to access and manage the system (Pazos 2005). Since, study of whole MetaRouter (www.pdg.cnb.uam.es/biodeg net/MetaRouter/) is freely available on the web.

7. Conclusion and Future Prospects

In order to gain a better knowledge of bioremediation, modern genetic molecular biology approaches, viz: molecular sequencing, genetic engineering, transcriptomics, proteomics, and interactomics, hold a lot of potential as tools for studying the mechanisms that control various metabolic pathways. Since the implications of these approaches are still in very far regarding the massive amounts of data created by recent genomics and proteomics technocrats require gradual organisation within informative databases. To comprehend the metabolic process in a precise way, technologies that combine transcriptomics and proteomics data must be refined. These methods have a lot of potential in terms of predicting organism metabolism in polluted environments and predicting microbial assisted attenuation of pollutants to speed up bioremediation. Future microbe-based bioremediation research could concentrate on the following topics:

1. Examination regarding existing data to get new knowledge about bioremediation routes (mechanism).
2. Numerical data modelling and simulation, which necessitates the development of novel algorithms.
3. Protocol standardization for data collection, processing, repositioning, and transmission.
4. Development and characterization of new indicators (biomarkers).

These signs could aid in deciding whether or not to apply specific bioremediation techniques. 5. Combination of data obtained through a variety of "omics" techniques.

Researchers may be able to obtain a better whole knowledge of microbe-based bioremediation pathways by combining community genomics (functional), transcriptomics, bioinformatics, as well as metabonomics data, and a complete, accurate perspective on the mixed cultures of coalition of microbes being required as the operational biodegradation procedure.

References

Abou Seeda, M. A., Yassen, A. A., and Abou El-Nour, E. Z. A. A. 2017. "Microorganism as a tool of bioremediation technology for cleaning waste and industrial water". *Biosci. Res.* 14, 633–636. DOI: https://doi.org/10.3389/fenvc.2020.570326.

Attila, C., Ueda, A., Cirillo, S. L. G., Cirillo, J. D., Chen, W., Wood, T. K. 2008. "*Pseudomonas aeruginosa* PAO1 virulence factors and poplar tree response in the Rhizosphere". *Microb Biotechnol* 2008, 1:17-29. DOI: 10.1111/j.1751-7915.2007.000 02.x.

Barupal, T., Chittora, D., and Meena, M. 2019. "Solid waste: characterization, assessment, monitoring, and remediation," in *Advances in Waste-to-Energy Technologies* DOI: https://doi.org/10.3389/fenvc.2020.570326.

Barupal, T., Meena, M., and Sharma, K. 2019. "Inhibitory effects of leaf extract of *Lawsonia inermis* on *Curvularia lunata* and characterization of novel inhibitory compounds by GC–MS analysis". *Biotechnol. Rep.* 23. DOI: 10.1016/j.btre.2019. e00335.

Benndorf, D., Balcke, G. U., Harms, H., von Bergen, M., 2007. "Functional metaproteome analysis of protein extracts from contaminated soil and groundwater". *ISME J.* 1, 224–234. DOI: 10.1038/ismej.2007.39.

Boltner, D., Godoy, P., Munoz-Rojas, J., Duque, E., Moreno-Morillas, S., Sanchez, L., Ramos, J. L. 2008. "Rhizoremediation of lindane by root colonizing Sphingomonas". *Microbial Biotechnol* 2008, 1:87-93. DOI: 10.1111/j.1751-7915.2007.00004.x.

Bursle, E., and Robson, J. 2016. "Non-culture methods for detecting infection". *Aust. Prescr.* 39 DOI: 10.18773/austprescr.2016.059.

Callaghan, A. V. 2013. "Metabolomic investigations of anaerobic hydrocarbon-impacted environments". *Curr. Opin. Biotechnol.* 24 DOI: 10.1016/j.copbio.2012.08.012.

Chakraborty, R., Wu, C. H., and Hazen, T. C. 2012. "Systems biology approach to bioremediation". *Curr. Opin. Biotechnol.* 23, 483–490. DOI: 10.1016/j.copbio.2012. 01.015.

Clarridge, J. E. 2004. "Impact of 16S rRNA gene sequence analysis for identification of bacteria on clinical microbiology and infectious diseases". *Clin. Microbiol. Rev.* 17, 840–862. DOI: 10.1128/CMR.17.4.840-862.2004.

Das, D., Mawlong, G. T., Sarki, Y. N., Singh, A. K., Chikkaputtaiah, C., and Boruah, H. P. D. 2020. "Transcriptome analysis of crude oil degrading *Pseudomonas aeruginosa* strains for identification of potential genes involved in crude oil degradation". *Gene* 755:144909. DOI: 10.1016/j.gene.2020.144909.

De Sousa, C. S., Hassan, S. S., Pinto, A. C., Silva, W. M., De Almeida, S. S., Soares, S. D. C., Azevedo, M. S.P., Rocha, C. S., Barh, D., Azevedo, V. "Microbial omics: applications in biotechnology," in *Omics Technologies and Bio-Engineering*, Eds D. Barh and V. Azevedo (Cambridge, MA: Academic Press), 3–20. DOI: 10.1016/B978-0-12-815870-8.00001-2.

Denef, V. J., Kalnejais, L. H., Mueller, R. S., Wilmes, P., Baker, B. J., Thomas, B. C., VerBerkmoes, Nathan C., Hettich, Robert L. and Banfield, Jillian F. 2010. "Proteogenomic basis for ecological divergence of closely related bacteria in natural acidophilic microbial communities". *Proc. Natl. Acad. Sci. U.S.A.* 107, 2383–2390. DOI: 10.1073/pnas.0907041107.

Dickson, R. P., Erb-Downward, J. R., Prescott, H. C., Martinez, F. J., Curtis, J. L., Lama, V. N., and Huffnagle, Gary B. 2014. "Analysis of culture-dependent versus culture-independent techniques for identification of bacteria in clinically obtained bronchoalveolar lavage fluid". *J. Clin. Microbiol.* 52, 3605–3613. DOI: 10.1128/JCM.01028-14.

Ellis, L. B. M., Roe, D., Wackett, L. P., 2006. "The University of Minnesota biocatalysis/biodegradation database: the first decade". *Nucleic Acids Res.* 34, D517–D521. DOI: 10.1093/nar/gkp771.

Espinosa, Cardenas M. J., Blanco, Colina A., Schmidgall, T., Atanasof-Kardjalief, A. K., Kappelmeyer, U., Tischler, D. 2020. "Toward biorecycling: isolation of a soil bacterium that grows on a polyurethane oligomer and monomer". *Front. Microbiol.* 11:404.doi: https://doi.org/10.3389/fmicb.2020.00404.

Gao, H., Wang, Y., Liu, X., Yan, T., Wu, L., Alm, E., Arkin, A., Thompson, D. K. and Zhou, J. 2004. "Global transcriptome analysis of the heat shock response of *Shewanella oneidensis.*" *J. Bacteriol.* 186, 7796–7803. DOI: 10.1128/JB.186.22.7796-7803.2004.

Gaytan, I., Sanchez-Reyes, A., Burelo, M., Vargas-Suarez, M., Liachko, I., Press, M., Sullivan, Sh., Cruz-Gómez, M. J. and Loza-Tavera, H. 2020. "Degradation of recalcitrant polyurethane and xenobiotic additives by a selected landfill microbial community and its biodegradative potential revealed by proximity ligation-based metagenomic analysis". *Front. Microbiol.* 10:2986. DOI: https://doi.org/10.3389/fmicb.2019.02986.

Gilbert, J. A., and Dupont, C. L. 2011. "Microbial metagenomics: beyond the genome". *Annu. Rev. Mar. Sci.* 3, 347–371. DOI: 10.1146/annurev-marine-120709-142811.

Gołebiewski, M., and Tretyn, A. 2020. "Generating amplicon reads for microbial community assessment with next generation sequencing". *J. Appl. Microbiol.* 128, 330–354. DOI: 10.1111/jam.14380.

Gupta, K., Biswas, R., and Sarkar, A. (2020). "Advancement of omics: prospects for bioremediation of contaminated soils," in *Microbial Bioremediation & Biodegradation*, ed M. Shah (Singapore: Springer), 113–42.

Han, M-J., Lee, S. Y. 2006. "The *Escherichia coli* proteome: past, present, and future prospects". *Microbiol Mol Biol Rev* 2006, 70:362-439. DOI: 10.1128/MMBR.00036-05.

Head, I. M., Singleton, I., and Milner, M. G. 2003. *Bioremediation: A Critical Review*. Wymondham: Horizon Scientific Press.

Holmes, D. E., O'Neil, R. A., Chavan, M. A., N'Guessan, L. A., Vrionis, H. A., Perpetua, L. A., Larrahondo, M. J., DiDonato, R., Liu, A., Lovely, D. R., 2009. Transcriptome of Geobacter uraniireducens growing in uranium-contaminated subsurface sediments. *ISME J.* 3, 216–230. DOI:https://doi.org/10.1038/ismej.2008.89.

Igiri, B. E., Okoduwa, S. I. R., Idoko, G. O., Akabuogu, E. P., Adeyi, A. O., and Ejiogu, I. K. 2018. "Toxicity and bioremediation of heavy metals contaminated ecosystem from tannery wastewater: a review". *J. Toxicol.* 2018:2568038. DOI: https://doi.org/10.1155/2018/2568038.

Jennings, L. K., Chartrand, M. M., Lacrampe-Couloume, G., Lollar, B. S., Spain, J. C., Gossett, J. M., 2009. Proteomic and transcriptomic analyses reveal genes upregulated by cis-dichloroethene in polaromonas JS666. Appl. Environ. Microbiol. 11, 3733–3744. DOI: 10.1128/AEM.00031-09.

Kapley, A., Prasad, S., and Purohit, H. J. 2007. "Changes in microbial diversity in fed-batch reactor operation with wastewater containing nitroaromatic residues". *Bioresour. Technol.* 98, 2479–2484. DOI: 10.1016/j.biortech.2006.09.012.

Kim, S. J., Jones, R. C., Cha, C. J. 2004. "Identification of proteins induced by polycyclic aromatic hydrocarbon in Mycobacterium vanbaalenii PYR-1 using two-dimensional polyacrylamide gel electrophoresis and de novo sequencing methods". *Proteomics* 4, 3899–3908. DOI: 10.1002/pmic.200400872.

Kim, Y. H., Cho, K., Yun, S. H., Kim, J. Y., Kwon, K. H., Yoo, J. S., Kim, S. I. 2006. "Analysis of aromatic catabolic pathways in Pseudomonas putida KT 2440 using a combined proteomic approach: 2DE/MS and cleavable isotope-coded affinity tag analysis". *Proteomics* 6, 1301–1318. DOI: 10.1002/pmic.200500329.

Kou, S., Vincent, G., Gonzalez, E., Pitre, F. E., Labrecque, M., and Brereton, N. J. B. 2018. "The response of a 16S ribosomal RNA gene fragment amplified community to lead, zinc, and copper pollution in a Shanghai field trial". *Front. Microbiol.* 9:366. DOI: https://doi.org/10.3389/fmicb.2018.00366.

Kuiper, I., Lagendijk, E. L., Bloemberg, G. V. Lugtenberg, BJJ. 2004. "Rhizoremediation: a beneficial plant–microbe interaction". *Mol Plant–Microbe Interact* 2004, 17:6-15. DOI: 10.1094/MPMI.2004.17.1.6.

Kumar, M., and Khanna, S. 2010. "Diversity of 16S rRNA and dioxygenase genes detected in coal-tar-contaminated site undergoing active bioremediation". *J. Appl. Microbiol.* 108, 1252–1262. DOI: https://doi.org/10.1111/j.1365-2672.2009.04523.x

Labie, D. 2007. "Developmental neurotoxicity of industrial chemicals". *Med. Sci.* 23, 868–872. DOI: 10.1051/medsci/20072310868.

Lau, H. T., Suh, H. W., Golkowski, M., and Ong, S. E. 2014. "Comparing SILACand stable isotope dimethyl-labeling approaches for quantitative proteomics". *J. Proteome Res.* 13, 4164–4174. DOI: 10.1021/pr500630a.

Liang, L., Stone, R. C., Stojadinovic, O., Ramirez, H., Pastar, I., Maione, A. G. 2016. "Integrative analysis of miRNA and mRNA paired expression profiling of primary fibroblast derived from diabetic foot ulcers reveals multiple impaired cellular functions". *Wound Repair Regen.* 24, 943–953. DOI: 10.1111/wrr.12470.

Lovely, D. R. 2003. "Cleaning up with genomics: applying molecular biology to bioremediation". *Nat. Rev. Microbiol.* 1, 35–44. DOI: https://doi.org/10.1038/nrmicro 731.

Lovely, D. R., Phillips, E. J. P., Gorby, Y. A., and Landa, E. R. 1991. "Microbial reduction of uranium". *Nature* 350, 413–416. DOI: https://doi.org/10.1038/350413a0.

Luciene, M. C., Helen, C. R., Luciana, M. C., Priscila, A. R., de Sousa Danielle, F. O. M., and Nívia, M. M. C. 2015. "Bioremediation of polluted waters using microorganisms," in *Advances in Bioremediation of Wastewater and Polluted Soil*, ed N. Shiomi (London: IntechOpen), 1–22. DOI: 10.5772/60770.

Malik, S., Beer, M., Megharaj, M., and Naidu, R. 2008. "The use of molecular techniques to characterize the microbial communities in contaminated soil and water". *Environ. Int.* 34, 265–276. DOI: 10.1016/j.envint.2007.09.001.

Mapelli, V., Olsson, Lisbeth, Nielsen, Jens. 2009. "Metabolic foot printing in microbiology: methods and applications in functional genomics and biotechnology". *Trends Biotechnol.* 26, 490–497. DOI: 10.1016/j.tibtech.2008.05.008.

Maroli, A, S., Gaines, T. A., Foley, M. E., Duke, S. O., Dogramacı, M., Anderson, J. V., Horvath, D. P., Chao, W. S. and Tharayil, N. 2018. "Omics in weed science: A perspective from genomics, transcriptomics, and metabolomics approaches". *Weed Sci.* 66, 681– 695. DOI: https://doi.org/10.1017/wsc.2018.33.

Matilla, M. A., Espinosa-Urgel, M., Rodriguez-Herva, J. J., Ramos, J. L., Ramos-Gonzalez, M. I. 2007. "Genomic analysis reveals the major driving forces of bacterial life in the rhizosphere". *Genome Biol* 2007, 8:R179. *DOI*: 10.1186/gb-2007-8-9-r179.

McGrath, K. C., Thomas-Hall, S. R., Cheng, C. T., Leo, L., Alexa, A., Schmidt, S., Schenk, P. M., 2008. "Isolation and analysis of mRNA from environmental microbial communities". *J. Microbiol. Methods* 75, 172–176. *DOI*: 10.1016/j.mimet.

Meena, M., Aamir, M., Vikas, K., Swapnil, P., and Upadhyay, R. S. 2018. "Evaluation of morpho-physiological growth parameters of tomato in response to Cd induced toxicity and characterization of metal sensitive NRAMP3 transporter protein". *Environ. Exp. Bot.* 148, 144–167. DOI:10.1016/j.envexpbot.2018.01.007.

Meena, M., Divyanshu, K., Kumar, S., Swapnil, P., Zehra, A., Shukla, V. 2019. "Regulation of L-proline biosynthesis, signal transduction, transport, accumulation and its vital role in plants during variable environmental conditions". *Heliyon* 5:e02951. DOI: 10.1016/j.heliyon.2019.e02952.

Misra, B. B., Langefeld, C. D., Olivier, M., and Cox, L. A. 2018. "Integrated omics: tools, advances, and future approaches". *J. Mol. Endocrinol.* 62, R21–R45. DOI: 10.1530/ JME-18-0055.

Nagalakshmi, U., Waern, K., and Snyder, M. 2010. "RNA-Seq: a method for comprehensive transcriptome analysis". *Curr. Protoc. Mol. Biol.* 4, 1–13. DOI: 10.1002/0471142 727.mb0411s89.

Nojiri, H., and Tsuda, M. 2005. "Functional evolution of bacteria in degradation of environmental pollutants". *Tanpakushitsu Kakusan Koso* 50, 1505–1509.

Okinaka, Y., Yang, C-H., Perna, N. T., Keen, N. T. 2002. "Microarray profiling of Erwinia chrysanthemi 3937 genes that are regulated during plant infection". *Mol Plant–Microbe Interac* 2002, 15:619-629. *DOI*: 10.1094/MPMI.2002.15.7.619.

Pandey, A., Tripathi, P. H., Tripathi, A. H., Pandey, S. C., and Gangola, S. 2019. "Omics technology to study bioremediation and respective enzymes," in *Smart Bioremediation Technologies: Microbial Enzymes*, ed. P. Bhatt (New Delhi: Academic Press), 23–43. DOI: https://doi.org/10.1016/B978-0-12-818307-6.00002-0.

Pantaleo, M. A., Ravegnini, G., Astolfi, A., Simeon, V., Nannini, M., Saponara, M. 2016. "Integrating miRNA and gene expression profiling analysis revealed regulatory networks in gastrointestinal stromal tumors". *Epigenomics* 8, 1347–1366. DOI: 10.2217/epi-2016-0030.

Paul, D., Pandey, G., Pandey, J., and Jain, R. K. 2005. "Accessing microbial diversity for bioremediation and environmental restoration". *Trends Biotechnol.* 23, 135–142. DOI: 10.1016/j.tibtech.2005.01.001.

Pazos, F., Guijas, D., Valencia, A., De Lorenzo, V., 2005. MetaRouter: bioinformatics for bioremediation. Nucleic Acids Res. 33, D588–D592. DOI: https://doi.org/10.1093/nar/gki068.

Pferdeort, V. A., Wood, T. K., Reardon, K. F. (2003). Proteomic changes in *Escherichia coli* TG1 after metabolic engineering for enhanced trichloroethylene biodegradation. *Proteomics* 2003, 3:1066-1069. DOI: 10.1002/pmic.200300418.

Pinel-Cabello M., Jroundi F., L, opez-Fern, andez, M., Geffers, R., Jarek, M., Jauregui, R., Link, A., Vílchez-Vargas, R. and Merroun, M. L. 2020. Multisystem combined uranium resistance mechanisms and bioremediation potential of *Stenotrophomonas bentonitica* BII-R7: Transcriptomics and microscopic study. *Journal of Hazardous Materials* 403. DOI:10.1016/j.jhazmat.2020.123858.

Poirier, I., Hammann, P., Kuhn, L., and Bertrand, M. (2013). Strategies developed by the marine bacterium *Pseudomonas fluorescens* BA3SM1 to resist metals: a proteome analysis. *Aquat. Toxicol.* 128, 215–232. DOI: 10.1016/j.aquatox.2012.12.006.

Pritchard, C. C., Cheng, H. H., and Tewari, M. (2012). MicroRNA profiling: approaches and considerations. *Nat. Rev. Genet.* 13, 358–369. DOI: 10.1038/nrg3198.

Raghunandan, K., Kumar, A., Kumar, S., Permaul, K., and Singh, S. (2018). Production of gellan gum, an exopolysaccharide, from biodiesel-derived waste glycerol by *Sphingomonas* spp. *3 Biotech.* 8:71. DOI: 10.1007/s13205-018-1096-3.

Rani, A., Porwal, S., Sharma, R., Kapley, A., Purohit, H. J., and Kalia, V. C. (2008). Assessment of microbial diversity in effluent treatment plants by culture dependent and culture independent approaches. *Bioresour. Technol.* 99, 7098–7107.DOI: 10.1016/j.biortech.2008.01.003.

Rogers, S. L., and McClure, N. (2003). "The role of microbiological studies in bioremediation process optimization," in *Bioremediation: A Critical Review*, eds I. M. Head, I. Singleton, and M. Milner (Norfolk: Horizon Scientific Press), 27–59.

Sakshi and Haritash, A. K. (2020). A comprehensive review of metabolic and genomic aspects of PAH-degradation. *Arch. Microbiol.* 202, 2033–2058. DOI: 10.1007/s0020 3-020-01929-5

Sato, Y., Hori, T., Koike, H., Navarro, R. R., Ogata, A., and Habe, H. (2019). Transcriptome analysis of activated sludge microbiomes reveals an unexpected role of minority nitrifiers in carbon metabolism. *Commun. Biol.* 2:179. DOI: https://doi.org/10.1038/s42003-019-0418-2.

Seo, J. S., Keum, Y. S., and Li, Q. X. (2009). Bacterial degradation of aromatic compounds. *Int. J. Environ. Res. Public Health* 6, 278–309. DOI: 10.3390/ijerph6010278.

Shendure, J. (2008). The beginning of the end for microarrays? *Nat. Methods* 5, 585–587. DOI: 10.1038/nmeth0708-585.

Tang, Y. J., Martin, H. G., Dehal, P. S., Deutschbauer, A., Llora, X., Meadows, A., Arkin, A., Keasling, J. D., 2009. Metabolic flux analysis of Shewanella spp. reveals evolutionary robustness in central carbon metabolism. Biotechnol. Bioeng. 102, 1161–1169. DOI: 10.1002/bit.22129.

Thakur, M., Medintz, I. L., and Walper, S. A. (2019). Enzymatic bioremediation of organophosphate compounds—progress and remaining challenges. *Front. Bioeng. Biotechnol.* 7, 289. DOI: https://doi.org/10.3389/fbioe.2019.00289.

Urich, T., Lanzen, A., Qi, J., Huson, D. H., Schleper, C., Schuster, F. S. C., 2008. Simultaneous assessment of soil microbial community structure and function through analysis of the meta-transcriptome. PLoS ONE 3 (e2527), 1–13. DOI: https://doi.org/10.1371/journal.pone.0002527.

Viant, M. R., and Sommer, U. 2013. "Mass spectrometry based environmental metabolomics: a primer and review". *Metabolomics* 9, 144–158. DOI: 10.1007/s11306-012-0412-x.

Wackett, L. P., Hershberger, C. D. 2001. "Biocatalysis and Biodegradation: Microbial Transformation of Organic Compounds Washington, D.C.": *ASM Press*; 2001.doi: http://www.asmscience.org/content/book/10.1128/9781555818036.

Wiechert, W., Schweissgut, O., Takanaga, H., Wolf B. Frommer, W. B., 2007. "Fluxomics: mass spectrometry versus quantitative imaging". *Curr. Opin. Plant Biol.* 10, 323–330. DOI: 10.1016/j.pbi.2007.04.015.

Wierckx, N. J. P., Ballerstedt, H., De Bont, J. A. M., DeWinde, J. H., Ruijssenaars, H. J., and Wery, J. 2008. "Transcriptome analysis of a phenol producing *Pseudomonas putida* S12 construct: genetic and physiological basis for improved production". *J. Bacteriol.* 190, 2822–2830. DOI: 10.1128/JB.01379-07.

Wierckx, N. J. P., Ruijssenaars, H. J., deWinde, J. H., Schmid, A., and Blank, L. M. 2009. "Metabolic flux analysis of a phenol producing mutant of *Pseudomonas putida* S12: verification and complementation of hypotheses derived from transcriptomics". *J. Biotechnol.* 143, 124–129. DOI: 10.1016/j.jbiotec.2009.06.023.

Wilmes, P., Wexler, M., Bond, P. L. 2008. "Metaproteomics provides functional insight into activated sludge wastewater treatment". *PLoS One* 12 (e1778), 1–11. DOI: https://doi.org/10.1371/journal.pone.0001778.

Wu, Y. R., Luo, Z. H., Kwok-Kei Chow, R., and Vrijmoed, L. L. P. 2010. "Purification and characterization of an extracellular laccase from the anthracene-degrading fungus *Fusarium solani* MAS2". *Bioresour. Technol.* 101, 9772–9777. DOI: 10.1016/j.biortech.2010.07.091.

Yao, C., Chen, B. H., Joehanes, R., Otlu, B., Zhang, X., Liu, Ch., Huan, T., Tastan, O., Cupples, L. A., Meigs, J. B., Fox, C. S., Freedman, J. E., Courchesne, P., O'Donnell, Ch. J., Munson, P. J., Keles, S. and Levy, D. 2015. "Integromic analysis of genetic variation and gene expression identifies networks for cardiovascular disease phenotypes". *Circulation* 131, 536–549. DOI: 10.1161/CIRCULATIONAHA.114.010696.

Yee, D. C., Maynard, J. A., Wood, T. K. 1998. "Rhizoremediation of trichloroethylene by a recombinant, root-colonizing *Pseudomonas fluorescens* strain expressing toluene orthomonooxygenase constitutively". *Appl Environ Microbiol* 1998, 64:112-118. DOI: 10.1128/AEM.64.1.112-118.1998.

Yoneda, A., Henson, W. R., Goldner, N. K., Park, K. J., Forsberg, K. J., Kim, S. J., Pesesky, M. W., Foston, M., Dantas, G., Moon, T. S. (2016). Comparative transcriptomics elucidates adaptive phenol tolerance and utilization in lipid-accumulating *Rhodococcus opacus* PD630. *Nucleic Acids Res.* 44, 2240–2254. DOI: 10.1093/nar/gkw055.

Zhong, M., and Zhou, Q. 2002. "Molecular-ecological technology of microorganisms and its application to research on environmental pollution". *Ying Yong Sheng Tai Xue Bao* 13, 247–251.

Zhuang, X., Han, Z., Bai, Z., Zhuang, G., and Shim, H. 2010. "Progress in decontamination by halophilic microorganisms in saline wastewater and soil". *Environ. Pollut.* 158, 1119–1126. DOI: 10.1016/j.envpol.2010.01.007.

Chapter 8

Transcriptome Advances: From Conventional Breeding to Genome-Assisted Breeding

Saima Shafi[1,*], Bisma Rashid[1], Malik Asif Aziz[1], Zaffar Mehdi Dar[1], Khalid Hussain Bhat[1], Mushtaq Ahmad Malik[1] and Owais Ali Wani[2]

[1]Division of Basic Sciences and Humanities, FOA, Wadura, Sopore, SKUAST-Kashmir, India

[2]Division of Soil Science, FOA, Wadura, Sopore, SKUAST-Kashmir, India

Abstract

Novel crops with higher production capacity, greater environmental resilience, and agro-biological characteristics that reduce chemical use are the need of the hour attributing to the market demand, climate change, and environmental concerns. Traditional breeding techniques can aid in the improvement of crop traits for instance that of improvement in quality, nutrition, as well as yield, but not at the necessary pace. They have been thoroughly studied to resolve the issue of curtailed yield in varied crops but haven't yielded any covet results. Next gen sequencing and through-put mechanism have approaches have unbolted vast research directions, also crop genomes have improved because of more profitable and advanced molecular breeding techniques. Sequencing-based methods, such as SAGE (Serial Study of Gene Expression), EST (Expressed sequence Tags Sequencing), MPSS (Massive Parallel Signature Sequencing), RNA-seq, and Hybridization-based approaches, such as Microarray Technology and Tiling-arrays, are some of the techniques widely used in plant breeding methods. Many transcriptomic

---

*Corresponding Author's E-mail: samshafi4@gmail.com.

In: Transcriptomics and their Importance
Editors: Sarah Naiyer and. Khalid Hussain Bhat
ISBN: 979-8-88697-044-9

methods such as DNA-Seq, RNA-Seq, ChIP-Seq, Methyl-Seq, MAINE-Seq, RIP-Seq and PhIT-Seq have advanced or vision about gene expression as well as gene regulation in plants. RNA-seq among these has emerged as the most efficient transcriptome tool. RNA-Seq also used to classify differentially expressed genes, has implications in functional genomics studies, particularly when the genomes being studied have limited knowledge. Crop development experiments focused on genomes and the obtainability of genomic data, as well as governing resources, tend to be unique to breeding programs. Many technological advances are being made which include expanded computing space, speedy processing, and improved investigative breakthrough and methods for downstream study and visualization. Because of the increased availability of genetic sequencing facilities and bioinformatics methods, analyses of individual genes are shifting to whole-genome research, which offers a more comprehensive view of how all genes interact. This could lead to a better sapience of gene functioning and potential regulatory issues within gene mesh, as well as more efficiently and quickly improved plants. The availability of plant genome sequences will assist with orphan crop research and use, as well as de novo domestication. In 21st century, "genomics-reinforced breeding" carries a lot of promise. In this chapter we discuss the transition from traditional breeding to modern genomic tools that will spur and quicken genomic-assisted breeding for the identification of avant-garde genes pertaining to increased yield and quality character resulting in long-term plant development.

Keywords: breeding, ChIP-Seq, DNA-Seq, expressed sequence tags (EST), genotypes, hybridization, massive parallel signature sequencing (MPSS), MAINE-Seq, microarray, RNA-seq, sequencing, Serial Study of Gene Expression (SAGE), transcriptomics, tiling-arrays

1. Introduction

Breeding has a censorious say in safeguarding food security since early 1900s and has a wrecking impact on food production throughout globe (Tester 2010; Shiferaw 2013). Nevertheless, from recent times, global food grade and quality matter have blown out as an out-turn of the unrestrained demand for food for increasingly growing population. Furthermore, extreme weather changes exacerbated by globally changing climate has led to heat as well as drought stress, resulting in major yield losses for farmers all over the world (Von Braun 2005). Planetary epidemics, such as the Irish Potato Blight during

1840s and the Southern corn leaf blight of 1970s in the United States, were catastrophic occurrences that killed a whole mass of people due unavailability of proper diet (Ristaino 2003; Tatum 1971). The proportion of food production to consumption has declined notably in recent time, but the global urbanization rates and population have considerably extended. People tend to eat processed foods, which have a lower nutritional quality, in this phase of rapid development and growth. By 2050, the world's population is projected to exceed to approximately 10 billion, but there aren't any proper plans in place to feed this overgrowing population (UN World., FAO, 2009). Developed countries agricultural production has enhanced, partly to meet the food requirements, while putting more pressure on food manufacturing departments (Voss-Fels 2019).

Breeding has always been exceptionally effective in cultivating crops varieties that have put up with the growth of modern societies since the very beginning of plant domestication around 10,000 years ago. Pre-genomics techniques have been used in the past. Since the middle of the twentieth century, scientific breeding methodologies have resulted in the production of new cultivars, which have helped in the up gradation of yield of most major crops. Plant breeding's success in the last century has been based on the efficient selection of desirable genetic combinations using appropriate breeding methods and the use of natural and mutant-induced genetic variation. The involvement of mutation-induced breeding in genetics have contributed to the development of a diversity of tools to classify selected genotypes for breeding purposes has been critical to plant breeding success. It also depends on the attestation of novel genes of interest (GOI) and the use of phenotype-based breeding methodologies (Perez 2012). Today, genomics provides new insight to the breeders with advanced generation methods and techniques allowing the analysis of the whole new genome, a paradigm change that allows for the unmediated genotype analysis and its relation to the phenotype (Tester 2010). Although classical genetics has transfigured breeding at the turn of twentieth century, genomics is ushering in new era of breeding in 21st Century.

Science related to genomics, as well as its use in plant breeding, is rapidly evolving. Intertwining of modern genomics with traditional crop cultivation has brought about a conventional change in the cropping technology resulting in modern genomics-based plant. Due to this exclusive breeding, transcriptomics plays a critical role for the development of highly evolved cultivars of crops which are highly in demand, as stated by FAO, for "greener revolution" required for feeding rising population while conserving the reservoirs. Major pillars of genomic advances are the advent of DNA seq technology, passively known as next generation sequencing (NGS) technology. These along with other technological advances gives breeders a genome-wide molecular resources (large sets of markers, high-density genetic maps, genotyping techniques new experiment population, and so on) that will be integrated incurrent breeding approaches (Varshney 2007a, b., Lorenz 2011). New plant breeding methodologies are emerging because of recent advances in genomics, which are enhancing and speeding up the molecular techniques in a variety of processes (e.g., association mapping, marker assisted selection, 'breeding by design,' gene pyramiding, genomic selection, and so on (Pelemann 2003; Collard 2008). Recent progression or implementation of molecular methods persisted towards breeding, especially for intricate traits, are presented and discussed in this study. We begin by introducing the most important genomic tools, followed by examples of how these tools can be used in breeding program. The purpose being providing modern breeders a comprehensive overview of how genomic assisted breeding could help in increasing performance of their breeding program.

2. Conventional Breeding

Since the beginning of cultivation, farmers have persistently tinkered with the genetic composition of the plantation. Early growers used to save the best quality crop, including the seeds for planting upcoming seasons. As the science of genetics began to succeed, the breeders started using what they learned about a plant's genome so as to be more specific in selecting desirable qualities to produce a more sophisticated variety. Wild relatives were less evolved rather than that of domesticated plant species due to various growth features, increased yields, pest infestation and diseases tolerance, size of seeds, or good quality fruits. Plant breeding has been practiced since a long time back and still is the most persistently accepted practice in present time period. Artificial mating or cross-pollination is done in crops to increase yield, as discovered in early times by different farmers. Due to progression of plant breeding, the breeders increased their knowledge about picking superior quality plants and crossing them to produce new and improved varieties of various crops.

Plant breeding as defined by E. Sanchez-Monge (Monge 1993) is "man's selection of the best plants as potential cultivars from a variable population." Neolithic man was a true plan, according to this description. Domestication syndrome induction in cultivated crops, which may have taken just a few decades, has phenotypically influenced our crop plants to the point that the earliest varieties known are very close to the plants we grow today. Crops, on the other hand, have evolved in tandem with human societies: as humans colonized new areas of the globe, their crops followed them. Crop species have developed into many local or landrace varieties due to their new condition's adaptability (Holden 1993). As a result of antrophic selection for yield and adaptation, a degree of genetic diversification greater than that found in natural populations of wild progenitors has been achieved (Blumler 1992). With the introduction of modern plant breeding, the expansion of intraspecific genetic diversity in cultivated plants came to an abrupt halt. The need for more specific metrics to use in gauging the deliberate improvement of agricultural plants was obvious long before the advent of genetics. Louis de Vilmoren was the first to use progeny research in the production of new varieties in the 18th century.

3. Emergence of New Breeding Paradigms: The Age of DNA Manipulation

Genetics and molecular biology has provided new ways to change the genetic material of living organisms in the last 30 years. It became relatively simple to elucidate gene structure and function since the advent of molecular cloning techniques. In comparison, how a single celled zygote differentiates into a full plant remains largely unclear in molecular terms. It's worth noting, however, that the gene, as we now know, contains the information needed for proper transcription and translation, both in terms of absolute expression and tissue specificity, in its coding unit. Recent advances in molecular biology have resulted in the creation of effective breeding methods, allowing the insertion of new genes isolated and manipulated in the lab into the plant genome, as well as changing their expression, especially when using Agrobacterium-based transformation vectors (Walden and Wingender, 1995). Plant-improvement genes are isolated from bacteria, cattle, plants, and other species, then manipulated before being reinserted into crop plants. Together, these modern approaches along with molecular improvement make up the biotechnology methodology. This ultimately allows for the development of living forms that have a low chance of arising naturally in the absence of human interference.

4. Transcriptomics

The transcriptome, is a vast field of molecular biology consisting of different transcripts of cell, tissue, or organism, including messenger RNA and noncoding RNAs (ncRNAs) (Morozova 2009). Transcriptomics compared with the genomic is, highly complicated and can vary based on different stages and reaction in response to environmental changes in given cell, tissue, or organ (with a few exceptions). Transcriptomics is employed in a variety of fields of sciences as it has vast implementation (Kliebestein 2012). By research into the transcriptomics researcher can now determine the gene of utmost importance and how they affect the working of genes. It now has become convenient to map transcripts onto genome giving various beneficial information which then will be exploited by the researchers (Wang, Z. Gerstein, and Snyder, 2009), resulting in the discovery of various gene allele and variants in vast number of species. More recently, several approaches for detecting and quantifying the transcriptome have been published. Next-gen seq (NGS) & microarray technologies are examples of high-throughput techniques which utilizes techniques such as the blotting i.e., northern blotting for detection of few transcripts to thousands of transcriptomes on basis of different genes present (Morozova 2009).

4.1. Studying the Transcriptome

A transcriptome can be studied using a variety of approaches and procedures. Sequencing-based technologies, such as EST sequencing, serial analysis of gene expression (SAGE), massive parallel signature sequencing (MPSS), RNA-seq, and hybridization-based approaches, such as microarray technology and tiling-arrays, can be split into two categories which are then used for study of transcriptome. Surveying a particular transcriptome result in the detection of the best among these methods which can be employed but these are based on various parameters, including the information from previously sequenced database, or other surveyed transcripts of various crops but it has its cost implications.

Genome sequencing is an essentiality for the determination of whole genome composition of an organism. A major objective for this plant genome study is the get an overview of the gene structure of crop and how the productivity of the crop will be increased which is the utmost objective of crop breeding. in genome sequencing a number of methods have been elucidate till date such as Targeted expressed gene sequencing, expressed sequence tags (EST), suppression subtractive hybridization (SSH) technique, to name a few. The function of these are quite different, like EST generated by sequencing methods provide quite a handful knowledge regarding the whole crop analysis. Also, they can serve as a directory for the gene sequences and help in comparative studying of various crops, and for the purpose of selection of desirable traits for agronomic practices (Bharti 2012). Likewise, Targeted expressed gene sequencing (TEGS) as the name suggest it focuses on a particular gene rather than the whole genome of crop.

The suppression subtractive hybridization (SSH) technique was also used to construct EST libraries, and ribosomal protein genes linked to cold and salt stress (Kim 2004). As far as related to study it was seen that, 478 drought-resilient ESTs were generated from tissues of the roots in of cicer arietinum (Buhariwalla 2005). Also, in another report by Deokar et al. in 2011, he isolated 3061 uni-genes from suppression subtractive hybridization libraries of underground as well as aboveground parts like the shoots and root tissues of drought-resistant cicer arietinum genotypes. Simiarly, suppression subtractive hybridization (SSH) was used to create 182 distinct sequences of ESTs from drought resistant and nonresistant pigeon-pea plants (Qiao 2012).

Table 1. ESTs present in plants as elucidated through Sanger sequencing in response to various stresses

Stress Caused	Crop	Sequenced ESTs
Drought	Garbano beans (Roots)	25,656
	Black Eyed Pea (Leaf, Stem and Meristem)	44,112
	Cowpea (Root)	17,148
	French Bean (Leaves)	6,201
	French Bean (Roots and aerial Parts)	7,563
	Soja Bean (Leaves)	5,247
	Soja Bean (Roots)	13,228
Hypersensitive response (HR) Induced.	Plant (9–11-day-old seedlings + the whole plant)	1,26,794
Wilt (Fusarium) and Mosaic Disease (Sterility Mosaic)	Angola Pea (Roots + Leaves)	9888
Aspergillus infection	Groundnut	826
Leaf spot & Tomato spot wilt virus		471
TSMV and Aspergillus parasiticus		16,618

Table 2. Some specific plant EST databases with their websites

Plant EST database	Website
Barley database (B-EST).	http://pgrc.ipk-gatersleben.de/est/est/login
Db-EST.	http://www.ncbi.nlm.nih.gov/dbEST/index.html
Database Pscroph.	http://www.pscroph.ucdavis.edu
ES Tree.	http://www.itb.cnr.it/estree
EST database Kazusa.	http://www.kazusa.or.jp/en/plant/database
E-GENES.	http://www.genome.jp/kegg-bin/create_kegg_
Genomics Network (Solanaceae).	http://sgn.cornell.edu/
Minnesota University.	http://www.ccgb.umn.edu/
Mendel-ESTS & Mendel-GFDb.	http://www.mendel.ac.uk/
Plant genome database.	http://www.zmdb.iastate.edu/PlantGDB/
Plant Gene Indices (TIGR).	http://www.tigr.org/tdb/tgi/plant.shtml
Resource Centre (Chlamydomonas).	http://www.biology.duke.edu/chlamy.genome
Rice EST Database.	http://www.ncpgr.cn
Sputnik.	http://www.mips.gsf.de/proj/sputni
Tbest DB.	http://www.tbestdb.bcm.umontreal.ca
Unigenes NCBI.	http://www.ncbi.nlm.nih.gov/UniGene/
VUS Mirror.	http://genome.cornell.edu/

Table 3. Latest research pertained to the study of RNA-sequencing and microarrays in some plants

Procedure	Crops	Study	Citation
Microarray	Canesugar	Gene expression profiles of sugarcane populations with varying sugar concentration	Felix et al., 2009
	Maize	Dry matter of maize and yield parameters	Fu et al., 2010
	Grape	Using a transcriptome atlas to study grape maturation	Fasoli et al., 2012
	Wheat	Probed molecular mechanisms cardinal to salinity tolerance of introgression line of wheat	Liu et al., 2012
	Cotton	Cotton fiber elongation mechanism	Gilbert et al., 2013
RNA-seq	Soyabean	Different plant tissues, and stages of growth and conditions of crop and how they are influenced with varying conditions	Libault et al., 2010
	Rice	Investigating the regulatory mechanisms that affect rice grain quality	Venu et al., 2011
	Sorghum	Investigating molecular mechanisms to examine the underlying osmotic stress	Dugas et al., 2011
	Citrus	Molecular mechanism involved in tolerance and susceptibility against pest Xylella fastidiosa	Rodrigues et al., 2013
	Tomato	Comparative study between transcriptome of domesticated and wild variety	Koenig et al., 2013
	Eucalyptus	Investigating genes and mechanism related to wood formation in three species of Eucalyptus	Salazar et al., 2013

Similarly, in case of ESTs, a greater number more than 3 million legume ESTs are currently in knowledge, mainly from glycine max. (1.5 million, Vodkin 2004), followed by the model legumes such as barrel clover (2,80,000), (Cheung 2006) and *lotus japonicas* (2,42 000), (Asamizu 2004). Within legume crop, cicer arietinum contributed around 200 000 ESTs (Muchero 2009), and faba bean 114 139 (Blair 2011). For salinity and drought stress, cDNA libraries have been generated from plants that suffer from these (Varshney 2009a) also cDNA libraries are developed in response to various diseases such as, wilt and sterility mosaic disease (SMD) (Raju 2010). ESTs compiled from stress receptive tissues through Sanger sequencing are listed in Table 1.

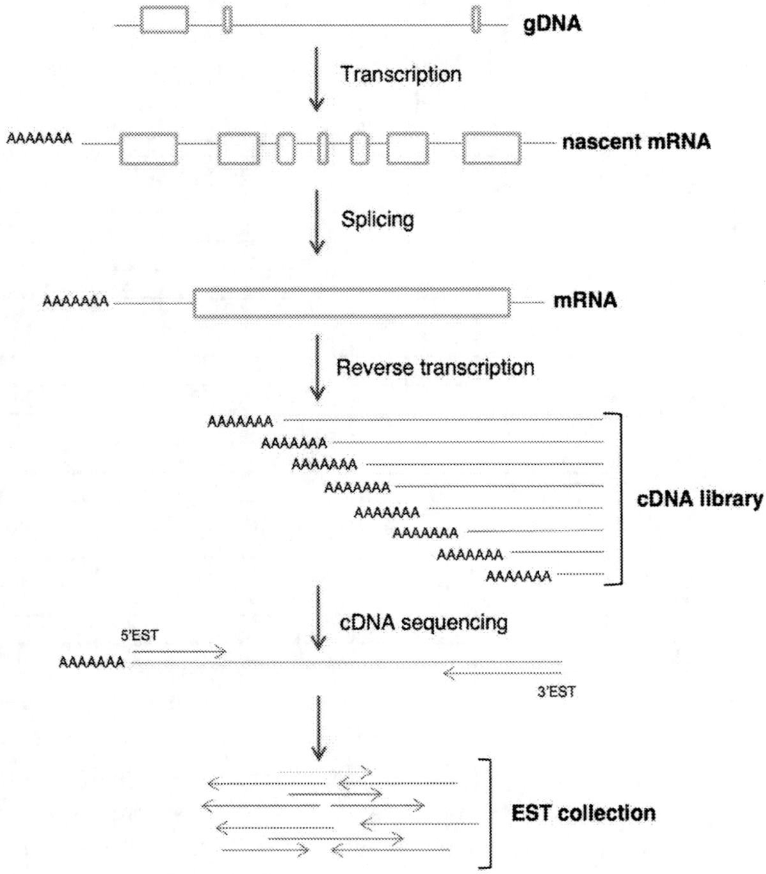

Figure 1. Overview of EST sequencing.

5. Applications of Transcriptomics Approaches for Crop Breeding

The majority of plant biology research is concentrated on identifying specific genes within a biological background since the early 2000s. Nowadays, researchers have made it quite an easy job to determine genes involved in working of plants and how they can be manipulated for better crop response (Morozova 2009). Expressive Sequence Tags are the first technique to be utilized for gene studying, discovery, and profiling of various crops. Many sequence tags have been elucidated in varying crops which are the model crops for studying, some of the examples being, oryza sativa, poplars and Arabidiopsis (Kumpatla and colleagues, 2012).

Table 4. ESTs submitted to NCBI pertaining to various crop plants (January 2013)

Crop	ESTs generated (dbEST)
Maize	21,19,138
Thale Cress	15,28,701
Soyabean	14,60,723
Wheat	12,85,373
Barley (Hordeum vulgare)	5,01,839
Rice (Oryza sativa)	12,53,558
Grapevine (Vitis vinifera)	4,46,665
Loblolly pine (Pinus taeda)	3,28,663
Apple (Malus Domestica)	32,503
Cotton (Gossypium hirsutum)	2,97,523
Tomato (Solanum lycopersicum)	2,97,143
Sorghum (Sorghum bicolor)	2,09,836
Orange (Citrus sinensis)	2,14,599
Coffee (Coffea arabica)	1,74,276
Cocoa (Theobroma cacao)	1,59,997
Sugarcane (Saccharum hybrid)	1,35,535
Populus (Populus trichocarpa)	89,944
Eucalyptus (Eucalyptus calmadulensis)	58,585

The NCBI database, provides an overview of over a million ESTs that are currently present with them. Over a 75 million ESTs are present in NCBI database as of January 2013 (http://www.ncbi.nlm.nih.gov/dbEST/), (Table 4). Several EST sequencing initiatives and EST databases have been established for agricultural species such as maize, sugarcane, and cotton (Table 5). Other databases have been built to provide EST collections and bioinformatics resources for EST mining, such as HarvEST (http://harvest.ucr.edu/) and PlantGDB (http://plantgdb.org/prj/) (Duvick 2008).

Table 5. EST sequencing projects and databases for crop species

Crop	EST Sequencing	Website
Malus pumila	Expressed Sequence Tags in Apple	http://titan.biotec.uiuc.edu/apple/
Citrus limon	CitEST	http://biotecnologia/centrodecitricultura.br
Coffee Arabica	Brazilian Coffee Genome Project	http://www.lge.ibi.unicamp/br/cafe/
Eucalyptus	Brazilian Eucalyptus Genome Sequence Project Consortium (FORESTs)	https://forests.esalq.usp.br/
Zea Mays	MGDP (Maize Gene Discovery Progress)	http://www.maizegdb.org/http://www.maizegdb.org/
Glycine Max	SoyBase	http://www.soybase.org/
Saccharum officinarum	Sugarcane EST Genome Project (SUCEST)b	http://sucest-fun.org/index.php/projects/sucest
Solanum lycopersicum	SOL Genomics Network (SGN)	http://solgenomics.net/
Triticum aestivum	International Triticeae EST Consortium (ITEC)	http://wheat.pw.usda.gov//genome/group/pool/

In many biological situations, ESTs have been utilized to identify a variety of transcripts. There are various ESTs that have been generate in response to various disease infected or resistant varieties (Cristofani-Yaly 2007), e.g., Citrus leprosis virus C (Freitas-Asta 2007), Xylellafastidiosa (De Souza 2007), & Citrus tristeza virus (Freitas-Asta 2007). Similarly, a varying number of ESTs are generated based on varying conditions in different crops (Kantety 2002), like, citrus (Palmieri 2007), or sugarcane (Vettore 2003), as well as single nucleotide polymorphism (SNP) markers from *Solanum lycopersiccon* (Yamamoto 2005), *Coffee Arabica* (Vidal 2010), and eucalyptus (Vi (Grattapaglia 2011). For many plant biologists, microarrays have become their preferred transcriptomic process. The Affymetrix ATH1 array platform was used to perform the most detailed microarray studies in Arabidopsis thaliana (Redman 2004). GENEVESTIGATOR, a database for microarray data mining, has huge information of Rockcress microarray that has created and introgressed (Zimmermann 2004). Likewise, a software, RiceXPro (Sato 2011) offers a medium for tracking rice genome in varying conditions of environment and how this stimulating environment affects the genome functioning (Benedito 2008 and Schmid 2005). Affymetrix developed and published arrays for different crop species as the ESTs count as well as genetic sequences increased, including the Affymetrix GeneChip® array for citrus, *Medicago spp., Gossypium, Hordium vulgarea, Triticum aestivum, Zea mays, populas, Orya sativa, Glycine max,* and *Solanum lycopersicon* (www.affymetrix.com). Nimble Gen/Agilent (http://www.agilent.com) is another organization that has developed and commercialized crop arrays. Thousands of transcripts have been profiled in resistant and sensitive citrus genotypes in response to pathogen infestation such as Candidatus Liberibacter spp., Xylellafastidiosa, and Xanthomonas spp., as well as the oomycete Phytophthoraparasitica, using microarray and, or recently, RNA-sequencing (Cernadas 2008; Albrecht and Bowman, 2012; Boava 2012; Mafra 2013; and Rodrigues 2013). Extensive study undertaken in citrus to clarify the susceptibility processes. Citrus limon breeding & genetically engineered commercial citrus cultivars that are more resistant to these infections have benefited from the transcriptome-identified target genes and molecular markers. Microarrays have also been employed to identify the molecular basis of complex features in different ways in conventional QTL research, linkages map is used to identify genetic areas associated to phenotypic difference (Kumpatla 2012). Using an approach known as "Genetic Genomics," RNA-sequencing and microarray approaces have been used to classify expressed QTLs (eQTLs) (Jansen and Nap, 2001). Plant adaptability and domestication can also be studied using microarrays (Koenig 2013). Microarrays were used in a recent study to look at how maize domestication has changed the transcriptome (Swanson-Wagner 2012). More recently, RNA sequencing has emerged as an uplifted tool in respect to microarray system as it gives complete feedback about the transcriptome resource for completely studied as well as less studied species (Varshney 2010). Like in some "orphan crops" like C*icer arietinum, Pennisetum glaucum, Ipomoea batatas* a large set of transcripts have been developed by using RNA sequencing. Likewise for more intricate gene composition, like *Triticum aestivum, Coffee Arabica* and *Saccharum*, RNA-seq has been successfully used as an alternative to whole genome sequencing (Pérez-de-Castro 2012). It has wide applications including its use in gene profiling and transcriptome discovery which in turn helps to determine the gene of interest. Its implementation in the discovery of SNPs also plays important roles. In eucalyptus 148 Mbp of ESTs were assembled using 454 technology which led to the discovery of 23000 SNPs. Same was used in triticum as well as Brasica crops (Trick 2009, 2012). RNA sequencing also provide an insight about noncoding (Morozova 2009) as well as MicroRNAs (miRNAs) Also it provides an insight about the impact of stimulating environment (Wang 2011; Curaba 2012) which can be used for crop breeding purposes.

5.1. Application of Transcriptomics in Elucidating Functional Genes of Interest

Identification for development and stress-related genes and pathways has been one of the most important applications of transcriptomics. In addition, various new applications have lately developed, including those that pertaine to molecular markers, and discovery of gene involved in secondary metabolism, including study of organ evolution in families.

5.2. Transcriptomic Studies Used in Elucidating Genes Involved in Crop Growth and Improvement

Due to successful application of transcriptomics in whole plant study, as well as in comparative studying, it has been proven to be a helping hand in recognizing the gene of utmost importance (Jia 2013; Nie M 2013) or how different genes react to changing environmental conditions (Huh 2013: Martzivanou 2003). The transcriptome of a gene family like C_2H_2 zinc finger or LRR reveals genes engaged in stress and how they affect the plant (Agarwal 2007: Zan 2013.). Whole transcriptome analysis reveals differentially expressed genes (DEGs), or preferentially expressed genes, variants, and novel genes, (Gao 2013; Jones 2013; Loraine 2013). Comparative study pertained to tissues or genomes provides an overview regarding various DEG (Cheng 2013; An 2014). Gene expression atlases can give a brief overview of regulatory processes entangled in spatio-temporal gene expression critical for vegetative growth as well as reproductive growth by generating comparison data from diverse topics and organs at the same time (Sharma 2014). Likewise intricate studies combining a certain cell, stage, organ, and character like quantitative analyses provides better feedback which are used for conducting various research (Gandotra 2013 Yu 2014). In apples, for example, the pigmentation is a valuable attribute that adds to the economic worth. In such cases, study of exocarps at varying developmental stages of the fruit in apple cv. red delicious fruits were analyzed using microarray technique which resulted in determining that the bHLH TFs & MYB are in charge for red apple pigmentation (Vimolmangkang 2014). Aside from that, the bigger picture is to link plant breeding with developmental research. Plant domestication, for example, is studied using gene expression analyses. RNA-seq investigation of ancient maize kernels revealed transcriptome alterations that may have happened throughout the domestication process (Guggisberg 2013). RNA-Sequencing data from 14 genotypes, an orthologs guided assembly technique was employed to produce a genome which was used as reference, with fragmentation and reduced transcript redundancy for diversity study in an out-breeding, highly hetero-zygogeneous plant, Loliumperenne (Ruttink 2013).

5.3. Improving Stress Tolerance with the Help of Transcriptome Data

Introduction of whole genome transcriptomics has boosted the field of transcriptome analysis that has aided in determining the stress repulsive genes and the processes, genes, pathways and or networks involved in avoidance of stress. Such information can be used to develop tolerant cultivars with higher yields. Transcriptome data on monocot crops (Shankar 2013; Zhang 2013), tree species (Chen 2012), edible plants such as soybean (Le 2012), and commercially important plants such as cotton (Yao 2011) have paved a new way for determining the ways to elucidate the harsh impact of stress on the crop. The connections between reproductive development, phyto-hormone signaling, and carbohydrate metabolism were investigated in a study of male gametophyte growth during drought stress (Jin 2013). Furthermore, a comparison of crop which is harshly affected by stress and a sensitive plant has shown the potential gene's modes of action on adaptation (Yang 2013). Different transcriptome analysis approaches working together can broaden the field for building a comparison platform to assess stress responses. To investigate responses to cold acclimation, RNA-seq data from a close relative of Arabidopsis was taken as a model for building microarray based on oligonucleotide sequence for specific expression profiling of genes (Lee 2013). The functional link of essential genes and transcription factors is highlighted by the effect of abiotic as well as biotic stress on different families, as well as on the co-expression analyses (Sarkar 2014; Wang 2014). The knowledge gained from these studies gave an outlook about the yield loss due to varying abiotic as well as biotic stresses and how to overcome those so as to avoid global food shortage.

5.4. Transcriptome of Non-Model Crops Act as Store House of Economically Important Information

NGS compared to microarray system has been proven to be more sophisticated and more dependable for studying of transcriptome of a non-model crop with higher economic feasibility. Various elements of non-model plant development and stress have been mobilized, in various crops including *Gossyply sps.* (cotton), "Suli" pear (Liu 2012), *Vitis Vinifera* (Sweetman 2012), *Glycine max* (Le 2012), Napa cabbage (Wang 2012), and Nettle spruges (Jiang 2012) to cite a few. Similarly, more particular facet of progress, such as evaluating different sexes in cucumber (Wu 2010), estimation of hormone status in apple trees (Krost 2012), and comparison of two cotton species base on different transcriptome (Lacape 2012) will allow the generation of economically valuable outcomes from the crops. Differentially Expressed Genes (DEGs) in Sweet Wormwood, i.e., *Artemisia annua*, during its growing stages of plant has shown differing levels for artemisia, an anti-malarial phytomolecule, that could be implicated in its manufacture (Nair 2013). In orchid, a Chinese medicinal plant, the transcriptome profiling, and gene characterization, have shown that there are various secondary important metabolites present in the plant that have a medicinal value (Bai 2013, Guo 2013).

Data on the manufacture of key chemicals in commercially relevant plants will lead to metabolic engineering research. As a result, genetically modified cultivars can be created by integrating or improving commercially relevant features such secondary metabolites with therapeutic value, with the aim being biodiesel feedstock, human-food products, essential- oils, and enhanced properties of ornamental plants. The potential of NGS is maximized by a shift in plant transcriptomics that lack genomic information despite being commercially important. The true "fruits" of this information will be realized through its application in plant breeding to produce superior cultivars.

5.5. Strength of Next Generation Sequencing through Small RNA Analysis

Next Gen Seq. is considered as one of the best methods to sequence short RNAs profile in various cells, tissues of both modal as well as non-modal plants (Kim 2012; Song 2013). Significance of the RNAs in regulating plant regulation has been studied using small RNA analysis (Lan 2012). There are many examples that have been studied in various crops like rice wherein, the role of miRNA was seen in rice male gametophyte development. Likewise in rose, same was seen wherein, profiling was done in 4 genotypes for identification of gene of interest affecting growth as well as desired phenotypes for improving floral features (Kim 2012). During times of environmental stress, small RNA sequencing can reveal a lot about the regulatory mechanisms that play a role in adaptation. Mangroves are excellent salt tolerant (halophies) & research into the transcriptome through short RNAs have revealed key genes associated with the salt tolerance (Khraiwesh 2013). Small RNAs, are predominantly derived from repeating region of plants (Farazi 2008). As a result, transcriptome research has focused on discovering molecular marker from the repeating portions of the loci that produce mature miRNAs (Joy 2013). Similarly small RNA profiling gives information on candidate gene regulation and the genes related to functional regulation. Marker knowledge and RNA profiling are very powerful tools for getting a know-how of regulation of genes at expressive and genomic levels and exploiting them for crop improvement.

5.6. Functional Genomic Studies in Plants Based on Transcriptome Approaches

There's a wealth of knowledge about transcriptomes of many plants in a variety of conditions. Genes have been identified and examined in this research for protecting the role of these in specific diseases or biological processes. Overexpression, mutations or using RNAi have been seen as an ultimate method to determine the regulatory pathways and down streaming genes involved in gene expression. Os-MYB3R-2, for example, discovered for its upregulation in rice (cold-stressed) & thus was proposed as key player. Transgenic Rockcress containing the same overly expressed genome construction portrayed a perfect example in tolerating stimulating environmental stressors such as extreme low temperatures, water scarcity and salt tolerance (Dai 2007). OsMYB2P-1, a member of the MYB family, was discovered through determination of phosphorus deficiency by microarray technique in *oryza sativa* and additional investigation revealed the role of this in root modification during the deficiency of phosphorous in roots (Dai 2012). Likewise for salt tolerance, in rockcress a cloned gene TaSRG from wheat salt tolerant variety was transcriptionally analyzed and was found as reliever for salt stress toleration, in Arabidopsis (Rockcress) crop (He 2011).

5.7. Transcriptomic Resources Used for Marker-Assisted Genetic Enhancement

Transcriptome tools are proving to be extremely useful in marker assisted breeding as well as for genome enhancement in various crops. Having the potential to create rapid marker-trait linkages as well as classify molecular tags which are functionally significant (novel genes, quantitative loci trait, genes and haplophyle types) regulating both quality and quantitative characteristics of agricultural significance (Varshney 2005). The large number of transcript sequences available for various crop genotypes, including controlled candidate, transcripts/genes would essentially help in generating large no. of wide-genome silicon polymorph genic Sequences Repeats (SSR), Single Nucleotide Polymorphisms markers (SNP) (Parida 2006; Parida 2008). Using high-throughput array-based NGS and marker genotyping, promising research have been done in validation and genotyping of individually accepted gene markers obtained from naturally diverse germplasm collection & population mapping of different crops (Gaur 2012). These recent innovations in the transcriptome analysis or evaluation of novel genomes/diversity have hiked the marker assisted breeding for the crop improvement. Now, complex quantitative traits for population mapping and determination of germplasm lines for multi environmental conditions can be done easily as there is quite an advancement in evaluating the phenotypic marker assisted breeding for modern phenotyping and E- typing platforms that can be used for the same (Clark 2011; Xu 2012). Likewise, for QTLs (trait-regulatory genes) separation in crops, high-resolution genic Single Sequence Repeats (SSRs) and single nucleotide polymorphisms (SNPs) marker-based transcript maps are one of the best in the field (K. Miura 2011). Genetical genomics, which combines these mapping of QTLs with profiling differential expressions is a quite established and effective method for cis and trans regulated expression identification. Lately, genome-wide association (GWAS) studies and gene-specific have become increasingly popular for potential trait-influencing genomic loci identification (genes, alleles, and QTLs) for genomic improvement of crop since the introduction of Next Gen Seq based whole genome global transcript sequence and high-throughput Single sequence repeats and Single Nucleotide Polymorphic marker genotyping assays (Y. Li 2011). By combining traditional breeding markers like MAS with that of more recently developed markers like multi-parent advanced generation intercross (MAGIC), genetic improvement of crops has been progressed lately (Septiningsih 2009). As a result, transcriptomic methods hold a lot of promise for speeding up genomics-assisted breeding in the production of genetically modified high-yielding climate-sensible agricultural varieties.

Conclusion and Future Prospects

Genes discovery for economically valuable features has changed the paradigm of what could compliment the ancient traditional cropping methods which could aid in the upliftment of the breeding process. Within these many enhanced transcript techniques and tools have been exploited as well as designed that has aided in the primary large-scale method for screening of various crops for economically important transcriptomes. Similarly, the development of a complete transcriptome, on the other hand, necessitates accurate transcript recognition and quantification using impartial, highly reliable, and cheaper or more economically feasible methods. RNA-seq among these transcriptomics has been elucidated as one of the most effectively exploited tool. As an evolving technology, however, many advancements are being made, including increased storage space, expeditious, and improved algorithms based on analytical methods and software for downstream analyses and visualization. In conclusion, more advanced methods, can be used to anatomize the genetic, physiological, biochemical, and evolutionary basis of complex traits by combining transcriptomic and other “omics” data. The great hope of “genomics-assisted breeding” in the twenty-first century is this method. If the population grows, likelihood of a food deprivation grows as well. Efforts to develop higher yielding cultivars that can withstand stimulating environmental distress must be bolstered. Traditional breeding techniques can aid in the improvement of different crop traits such as quality, nutritive value, and yield, but not at what is required. Scientists are using more thrifty and improved molecular methods of breeding to improve the genomes of crops. More recently, NGS & RNA sequencing are used as a tool for gene identification, construction of gene mapping, which help to improve qualitative and quantitative traits in many crops. These improved technologies would aid in the production and advancement of various crops and breeding programs, while also being cost-effective, environmentally sustainable, and, most importantly, time efficient.

References

Adams, M.D., Kelley, J.M. and Gocayne, J.D. 1991. "Complementary DNA sequencing: expressed sequence tags and human genome project." *Science* 252 (5013): 1651–1656. doi:10.1126/science.2047873.

Agarwal, P., Arora, R., Ray, S. and Singh, A.K. 2007. "Genome-wide identification of C2H2 zinc-finger gene family in rice and their phylogeny and expression analysis." *Plant Molecular Biology* 65, 467–485. doi: 10.1007/s11103-007-9199-y.

Aishwarya, V. and Sharma, P.C. 2008. "UgMicroSatdb: Database for mining microsatellites from unigenes." *Nucleic Acids Research* 36, D53– D56. doi: 10.1093/nar/gkm811.

Albrecht, U. and Bowman, K.D. 2012. "Transcriptional response of susceptible and tolerant citrus to infection with *Candidatus Liberibacter asiaticus. Plant Science* 185: 118–130. doi: 10.1016/j.plantsci.2011.09.008 .

An, H., Yang, Z., Yi, B. and Wen, J. 2014. "Comparative transcript profiling of the fertile and sterile flower buds of pol CMS in *B. napus*." *BMC Genomics* 15, 258. doi: 10.1186/1471-2164-15-258.

Anders, A. and Huber, W. 2010. "Differential expression analysis for sequence count data." *Genome Biology* 11 (10): R106. doi.org/10.1186/gb-2010-11-10-r106.

Anders, S., Pyl, P.T. and Huber, W. 2015. "HTSeq—a python framework to work with high-throughput sequencing data." *Bioinformatics* 31:166–169. Doi: doi.org/10.1093/bioinformatics/btu638.

Asamizu, E., Nakamura, Y., Sato, S. and Tabata, S. 2004. "Characteristics of the Lotus japonicus gene repertoire deduced from large-scale expressed sequence tag (EST) analysis." *Plant Molecular Biology* 54, 405–414. doi: doi.org/10.1023/B:PLAN.0000036372.46942.b8.

Au, K.F., Jiang, H., Lin, L., Xing, Y. and Wong, W.H. 2010. "Detection of splice junctions from paired-end RNA-seq data by SpliceMap." *Nucleic Acids Research*, 38 (14): 4570–4578. doi: 10.1093/nar/gkq211.

Bai, S., Saito, T., Sakamoto, D., and Ito, A. 2013. "Transcriptome analysis of Japanese pear (Pyrus pyrifolia Nakai) flower buds transitioning through endodormancy." *Plant Cell Physiology* 54, 1132–1151. doi: doi.org/10.1093/pcp/pct067.

Benedito, V.A., Torres-Jerez, I. and Murray, J.D. 2008. "A gene expression atlas of the model legume Medicago truncatula." *The Plant Journal* 55 (3): 504–513 doi:10.1111/j.1365-313X.2008.03519.x.

Blumler, M.A. 1992. "Independent Inventionism and recent genetic advance on plant domestication." *Economic Botany* 46: 98-111.

Boava. L., Cristofani-Yaly, M. and Mafra, V. 2011. "Global gene expression of Poncirus trifoliata, Citrus sunki and their hybrids under infection of Phytophthora parasitica." *BMC Genomics* 12 (1): 39. doi: https://doi.org/10.1186/1471-2164-12-39.

Bray, N.L., Pimentel, H., Melsted, P. and Pachter, L. 2016. "Near-optimal probabilistic RNA-seq quantification." *Nature Biotechnology* 34:525–527. doi: https://doi.org/10.1038/nbt.3519.

Brown, O.P. 1995. "Quantitative monitoring of gene expression patterns with a complementary DNA microarray." *Science*, 270: 467–470. doi: doi.org/10.1023/B:JOCI.0000025443.44833.1d.

Buhariwalla, H.K., Jayashree, B., Eshwar, K. and Crouch, J.H. 2005. "Development of ESTs from chickpea roots and their use in diversity analysis of the Cicer genus." *BMC Plant Biology* 5(1), 16. Doi: https://doi.org/10.1186/1471-2229-5-16.

Cheung, F., Haas, B.J., Goldberg, S.M.D., May, G.D., Xiao, Y. and Town, C.D.2006. "Sequencing Medicago truncatula expressed sequenced tags using 454 Life Sciences technology." *BMC Genomics* 7, 272. Doi: doi.org/10.1071/FPv40n12_FO.

Curaba, J., Spriggs, A. and Taylor, J. 2012. "miRNA regulation in the early development of barley seed." *BMC Plant Biology* 12 (1): 120. doi: https://doi.org/10.1186/1471-2229-12-120.

Cristofani-Yaly, M., Berger, I.J. and Targon, M.L.P.N. 2007. "Differential expression of genes identified from Poncirus trifoliata tissue inoculated with CTV through EST analysis and in silico hybridization." *Genetics and Molecular Biology*, 30 (3): 972–979. doi: https://doi.org/10.1590/S1415-47572007000500025.

Cernadas, R.A., Camillo, L.R. and Benedetti, C.E. 2008. "Transcriptional analysis of the sweet orange interaction with the citrus canker pathogens *Xanthomonas axonopodi* spv. *citri* and *Xanthomonas axonopodispv. aurantifolii*." *Molecular Plant Pathology* 9 (5): 609–631. doi: 10.1111/j.1364-3703.2008.00486.x.

Chen, S., Jiang, J., Li, H. and Liu, G. 2012. "The salt-responsive transcriptome of Populus simonii × Populus nigra via DGE." *Gene*, 504, 203–212. doi: 10.1093/jisesa/iex050.

Cheng, L., Li, S., Xu, X. and Hussain, J. 2013. Identification of differentially expressed genes relevant to corm formation in Sagittaria trifolia. *PLoS ONE,* 8, e54573.doi: doi.org/10.1371/journal.pone.0054573.

Collard, B.C.Y. and Mackill, D.J. 2008. "Marker-assisted selection: an approach for precision plant breeding in the twenty-first century." *Phil. Transactions Royal Society* 363, 557-572. doi: https://doi.org/10.1098/rstb.2007.2170.

Clark, R.T., MacCurdy, R.B., Jung, J.K. and Shaff, J.E. 2011. "Threedimensional root phenotyping with a novel imaging and software platform." *Plant Physiology* 156, 455–465. doi: 10.1104/pp.110.169102.

Dai, X., Xu, Y., Ma, Q. and Xu, W. 2007. "Overexpression of an R1R2R3 MYB gene, OsMYB3R-2, increases tolerance to freezing, drought, and salt stress in transgenic Arabidopsis." *Plant Physiology* 143, 1739–1751. doi: https://doi.org/10.1104/pp.106.094532.

Dai, X., Wang, Y., Yang, A. and Zhang, W.H. 2012. "OsMYB2P-1, an R2R3 MYB transcription factor, is involved in the regulation of phosphate-starvation responses and root architecture in rice." *Plant Physiology* 159, 169–183. doi: 10.1104/pp.112.194217 .

De Souza, A.A., Takita, M.A. and Coletta-Filho. 2007. "Comparative analysis of differentially expressed sequence tags of sweet orange and mandarin infected with Xylella fastidiosa." *Genetics and Molecular Biology* 30 (3): 965–971. doi: https://doi.org/10.1186/1471-2164-14-676.

Deokar, A.A., Kondawar, V., Jain, P.K., Karuppayil, S.M., Raju, N.L., Vadez, V., Varshney, R.K. and Srinivasan, R. 2011. "Comparative analysis of expressed sequence tags (ESTs) between drought-tolerant and-susceptible genotypes of chickpea under terminal drought stress." *BMC Plant Biology* 11(1), 70. doi: https://doi.org/10.1186/1471-2229-11-70.

Desai, S., Hili, J. and Trelogan, S. 2001. "Identification of differentially expressed genes by suppression subtractive hybridization. In: Stephen, H. and Levesey, R. (eds)." *Functional Genomics: A Practical Approach. Oxford: Oxford, University Press*, pp. 45–80. doi:10.1002/pmic.201200275.

Diatchenko, L., Lau, Y.F. and Campbell, A.P. 1996. "Suppression subtractive hybridization: a method for generating differentially regulated or tissue-specific cDNA probes and libraries." *Proceedings of the National Academy of Sciences* U.S.A. 93 (12): 6025–6030. doi: https://doi.org/10.1073/pnas.93.12.6025.

Duvick, J., Fu, A. and Muppirala, U. 2008. "PlanrtGDB: a resource for comparative plant genomics" *Nucleic Acids Research* 36: D959–D965. doi: https://doi.org/10.1093/nar/gkm1041.

Dugas, D.V., Monaco, M.K. and Olson, A. 2011. "Functional annotation of the transcriptome of Sorghum bicolor in response to osmotic stress and abscisic acid." *BMC Genomics* 12 (1): 514. doi: https://doi.org/10.1186/1471-2164-12-514.

Edwards, D., Batley, J. and Snowdon, R.J. 2013. "Accessing complex crop genomes with next-generation sequencing." *Theoretical and Applied Genetics* 126 (1), 1–11 doi: 10.1007/s00122-012-1964-x .

FAO. *How to Feed the World in 2050. Insights from an Expert Meet.* FAO 2009, 2050, 1–35.

Farazi, T.A., Juranek, S.A. and Tuschl, T. 2008. "The growing catalog of small RNAs and their association with distinct Argonaute/Piwi family members." *Development*, 135, 1201–1214. doi: 10.1242/dev.005629.

Fasoli, M., Dal Santo, S. and Zenoni, S. 2012. "The grapevine expression atlas reveals a deep transcriptome shift driving the entire plant into a maturation program." *The Plant Cell Online* 24 (9): 3489–3505. doi: 10.1105/tpc.112.100230.

Felix, J.M., Papini-Terzi, F.S. and Rocha, F.R. 2009. "Expression profile of signal transduction components in a sugarcane population segregating for sugar content." *Tropical Plant Biology* 2: 98–109. doi:10.1007/s12042-009-9031-8.

Freitas-Astúa, J., Bastianel, M. and Locali-Fabris, E.C. 2007. "Differentially expressed stress-related genes in the compatible citrus-Citrus leprosis virus interaction." *Genetics and Molecular Biology* 30: 980–990. doi:10.1590/S1415-47572007000500026.

Fu, J., Thiemann, A. and Schrag, T. 2010. "Dissecting grain yield pathways and their interactions with grain dry matter content by a two-step correlation approach with maize seedling transcriptome." *BMC Plant Biology* 10 (1): 63. doi: https://doi.org/10.1186/1471-2229-10-63.

Fu, X., Fu, N. and Guo, S. 2009. "Estimating accuracy of RNA-Seq and microarrays with proteomics." *BMC Genomics* 10 (1): 161. doi: 10.1186/1471-2164-10-161 .

Gaur, R., Azam, S., Jeena, G. and Khan, A.W. 2012. "High-throughput SNP discovery and genotyping for constructing a saturated linkage map of chickpea (*Cicer arietinum* L.)." *DNA Research* 19, 357–373. doi: 10.1186/1471-2229-13-161.

Gao, Y., Xu, H., Shen, Y. and Wang, J. 2013. "Transcriptomic analysis of rice (Oryza sativa) endosperm using the RNA-Seq technique." *Plant Molecular Biology* 81, 363–378. doi: 10.1186/s12918-018-0574-x.

Gandotra, N., Coughlan, S.J. and Nelson, T. 2013. "The Arabidopsis leaf provascular cell transcriptome is enriched in genes with roles in vein patterning." *Plant Journal* 74, 48–58. doi: 10.1007/s00299-018-2319-0.

Guggisberg, A., Lai, Z., Huang, J. and Rieseberg, L.H. 2013. "Transcriptome divergence between introduced and native populations of Canada thistle, *Cirsium arvense*." *New Phytology* 199, 595–608. doi: 10.1111/nph.12258 .

Gilbert, M.K., Turley, R.B. and Kim, H.J. 2013. "Transcript profiling by microarray and marker analysis of the short cotton (*Gossypium hirsutum* L.) fiber mutant Ligon lintless-1 (Li1)." *BMC Genomics* 14 (1): 403. doi: 10.1186/1471-2164-14-403 .

Grabherr, M.G., Haas, B.J. and Yassour, M. 2011. "Full-length transcriptome assembly from RNA-Seq data without a reference genome." *Nature Biotechnology* 29 (7): 644–652. doi: https://doi.org/10.1038/nbt.1883.

Griffith, M., Griffith, O.L. and Mwenifumbo, J. 2010. "Alternative expression analysis by RNA sequencing." *Nature Methods* 7 (10): 843–847. doi: 10.1038/nmeth.1503 .

Guo, X., Li, Y., Li, C. and Luo, H. 2013. "Analysis of the Dendrobium officinale transcriptome reveals putative alkaloid biosynthetic genes and genetic markers." *Gene* 527, 131–138. doi: 10.1016/j.gene.2013.05.073.

Grabherr, M.G., Haas, B.J. and Yassour, M. 2011. "Full-length transcriptome assembly from RNA-Seq data without a reference genome." *Nature Biotechnology* 29 (7): 644–652. doi: https://doi.org/10.1038/nbt.1883.

Grattapaglia, D., Silva-Junior, O.B., Kirst, M. 2011. "High-throughput SNP genotyping in the highly heterozygous genome of Eucalyptus: assay success, polymorphism and transferability across species." *BMC Plant Biology* 11 (1): 65. doi: https://doi.org/10.1186/1471-2229-11-65.

He, X., Hou, X., Shen, Y., and Huang, Z. 2011. "TaSRG, a wheat transcription factor, significantly affects salt tolerance in transgenic rice and Arabidopsis." FEBS Lett, 585, 1231–1237. doi: https://doi.org/10.1016/j.febslet.2011.03.055.

Holden, J., J. Peacock and T. Williams. 1993. "Genes, Crops and the Environment." University Press, Cambridge. The journal of Agricultural Science 125 (1), pp. 165 doi: https://doi.org/10.1017/S0021859600074621.

Huh, S.M., Hwang, Y.S., Shin, Y.S. and Nam, M.H. 2013. "Comparative transcriptome profiling of developing caryopses from two rice cultivars with differential dormancy." *Journal of Plant Physiology* 170, 1090– 1100. doi: https://doi.org/10.3390/genes 12122013.

Jansen, R.C. and Nap, J.P. 2001. "Genetical genomics: the added value from segregation." *Trends in Genetics* 17 (7): 388–391. doi: 10.1016/s0168-9525(01)02310-1.

Jiao, X., Zhao, X., Zhou, X.R., and Green, A.G. 2013. "Comparative transcriptomic analysis of developing cotton cotyledons and embryo axis." PLoS ONE, 8, e71756. doi: https://doi.org/10.1371/journal.pone.0071756.

Jiang, H., Wu, P., Zhang, S. and Song, C. 2012. "Global analysis of gene expression profiles in developing physic nut (Jatropha curcas L.) seeds." PLoS ONE, 7, e36522. doi: https://doi.org/10.1371/journal.pone.0036522.

Jin, Y., Yang, H., Wei, Z., Ma, H. and Ge, X. 2013. "Rice male development under drought stress: Phenotypic changes and stage-dependent transcriptomic reprogramming." *Molecular Plant* 6, 1630–1645. doi: https://doi.org/10.1093/mp/sst067.

Jones, S.I. and Vodkin, L.O. 2013. "Using RNA-Seq to profile soybean seed development from fertilization to maturity." PLoS ONE, 8, e59270. doi: https://doi.org/10.1371/ journal.pone.0059270.

Joy, N., Asha, S., Mallika, V. and Soniya, E.V. 2013. "De novo transcriptome sequencing reveals a considerable bias in the incidence of simple sequence repeats towards the downstream of 'Pre-miRNAs' of black pepper." PLoS ONE, 8, e56694.doi: https://doi.org/10.1371/journal.pone.0059270.

Khraiwesh, B., Pugalenthi, G. and Fedoroff, N.V. 2013. "Identification and analysis of red sea mangrove (Avicennia marina) microRNAs by high-throughput sequencing and their association with stress responses." PLoS ONE, 8, e60774. doi: https://doi.org/ 10.1371/journal.pone.0060774.

Kim, K.Y., Park, S.W., Chung, Y.S., Chung, C.H., Kim, J.I. and Lee, J.H. 2004. "Molecular cloning of low-temperature-inducible ribosomal proteins from soybean." *Journal of Experimental Botany* 55, 1153–1155. doi: 10.1093/jxb/erh125.

Kim, J., Park, J.H., Lim, C.J. and Lim, J.Y. 2012. "Small RNA and transcriptome deep sequencing proffers insight into floral gene regulation in Rosa cultivars." *BMC Genomics,* 13, 657. doi: https://doi.org/10.1186/1471-2164-13-657.

Kliebenstein, D.J. 2012. "Exploring the shallow end; estimating information content in transcriptomics studies." *Frontiers in Plant Science*, 3: 213.

Koenig, D., Jiménez-Gómez, J.M. and Kimura, S. 2013. "Comparative transcriptomics reveals patterns of selection in domesticated and wild tomato." *Proceedings of the National Academy of Sciences* U.S.A., 110 (28): E2655–E2662. doi: 10.1073/pnas.130 9606110.

Krost, C., Petersen, R. and Schmidt, E.R. 2012. "The transcriptomes of columnar and standard type apple trees (Malus x domestica) – a comparative study." *Gene*, 498, 223–230. doi: 10.1016/j.gene.2012.01.078.

Kantety, R.V., La Rota, M., Matthews, D.E. and Sorrells, M.E. 2002. "Data mining for simple sequence repeats in expressed sequence tags from barley, maize, rice, sorghum and wheat." *Plant Molecular Biology* 48 (5-6): 501–510. doi: 10.1023/a:10148752061 65.

Kumpatla, S.P., Buyyarapu, R., Abdurakhmonov, I.Y. and Mammadov, J.A. 2012. "Genomics-assisted plant breeding in the 21st century: technological advances and progress." In: Abdurakhmonov, I. (ed.). *Plant Breeding*. Rijeka, Croatia: In Tech, pp. 131–184. doi:10.5772/1389.

Koenig, D., Jiménez-Gómez, J.M. and Kimura, S. 2013. "Comparative transcriptomics reveals patterns of selection in domesticated and wild tomato." *Proceedings of the National Academy of Sciences* U.S.A., 110 (28): E2655–E2662. doi: 10.1073/pnas. 1309606110.

Lan, Y., Su, N., Shen, Y. and Zhang, R. 2012. "Identification of novel miRNAs and miRNA expression profiling during grain development in indica rice." *BMC Genomics* 13, 264. doi: https://doi.org/10.1186/1471-2164-13-264.

Liao, Y., Smyth, G.K, and Shi, W. 2014." FeatureCounts: an efficient general purpose program for assigning sequence reads to genomic features." *Bioinformatics Oxford* England, 30:923–930. doi: 10.1093/bioinformatics/btt656.

Libault, M., Farmer, A. and Joshi, T.G. 2010. "An integrated transcriptome atlas of the crop model Glycine max, and its use in comparative analyses in plants." *The Plant Journal*, 63 (1): 86–99. doi: https://doi.org/10.1071/FP13191.

Liu, C., Li, S., Wang, M. and Xia, G. 2012. "A transcriptomic analysis reveals the nature of salinity tolerance of a wheat introgression line." *Plant Molecular Biology* 78 (1-2): 159–169. doi:10.1007/s11103-011-9854-1.

Li, Y., Fan, C., Xing, Y. and Jiang, Y. 2011. "Natural variation in GS5 plays an important role in regulating grain size and yield in rice." *Nature Genetic* 43, 1266–1269. doi: https://doi.org/10.1016/j.molp.2017.03.009.

Lorenz, A.J., Chao, S., Asoro, F.G., Heffner, E.L., Hayashi, T., Iwata, H., Smith, K.P., Sorrells, M.K. and Jannink, J.L. 2011. "Genomic selection in plant breeding: knowledge and prospects." *Advamced Agronomy* 110, 77-123. doi: https://doi.org/10. 1590/1984-70332021v21Sa28.

Love, M.I., Huber, W. and Anders, S. 2014. "Moderated estimation of fold change and dispersion for RNA-seq data with DESeq2." *Genome Biology* 15. doi: https://doi.org/ 10.1186/s13059-014-0550-8.

Luk'ianov, S.A., Gurskaia, N.G. and Luk'ianov, K.A. 1994. "Highly-effective subtractive hybridization of cDNA." *Bioorganicheskaya khimiya* 20: 701–704.

Lukyanov, S.A., Rebrikov, D. and Buzdin, A.A. 2007. "Suppression subtractive hybridization." In: Buzdin, A.A. and Lukyanov S.A. (eds). Nucleic Acids Hybridization Modern Applications. Dordrecht, The Netherlands: *Springer*, pp. 53–84.

Le, D.T., Nishiyama, R., Watanabe, Y. and Tanaka, M. 2012. "Differential gene expression in soybean leaf tissues at late developmental stages under drought stress revealed by genome-wide transcriptome analysis." *PLoS ONE,* 7, e49522. doi: https://doi.org/10. 1371/journal.pone.0049522.

Lee, Y.P., Giorgi, F.M., Lohse, M. and vederaviciute, K. 2013. "Transcriptome sequencing and microarray design for functional genomics in the extremophile Arabidopsis relative *Thellungiella salsuginea* (*Eutrema salsugineum*)." *BMC Genomics* 14, 793. doi: https://doi.org/10.1186/1471-2164-14-793.

Liu, G., Li, W., Zheng, P. and Xu, T. 2012. "Transcriptomic analysis of 'Suli' pear (Pyrus pyrifolia white pear group) buds during the dormancy by RNA-Seq." *BMC Genomics*, 13, 700. doi: https://doi.org/10.1186/1471-2164-13-700.

Lacape, J.M., Claverie, M., Vidal, R.O. and Carazzolle, M.F. 2012. "Deep sequencing reveals differences in the transcriptional landscapes of fibers from two cultivated species of cotton." *PLoS ONE,* 7, e48855. doi: https://doi.org/10.1371/journal.pone.0048855.

Loraine, A.E., McCormick, S., Estrada, A., Patel, K. and Qin, P. 2013. "RNA-seq of Arabidopsis pollen uncovers novel transcription and alternative splicing." *Plant Physiology* 162, 1092–1109. doi: https://doi.org/10.1104/pp.112.211441.

Madden, S.L., Wang, C.J. and Landes, G. 2000. "Serial analysis of gene expression: from gene discovery to target identification." *Drug Discovery Today* 5 (9): 415–425. doi: 10.1016/s1359-6446(00)01544-0.

Matsumura, H., Reich, S. and Ito, A. 2003. "Gene expression analysis of plant host–pathogen interactions by SuperSAGE." Proceedings of the National Academy of Sciences U.S.A., 100: 15718–15723. doi: https://doi.org/10.1073/pnas.2536670100.

Matsumura, H., Nirasawa, S. and Terauchi, R. 1999. "Technical advance: transcript profiling in rice (*Oryza sativa* L.) seedlings using serial analysis of gene expression." *Plant Journal* 20: 719–726. doi: 10.1104/pp.105.060988.

Mockler, T.C., Chan, S. and Sundaresan, A. 2005. "Applications of DNA tiling arrays for whole-genome analysis." *Genomics*, 85: 1–15. doi: 10.1016/j.ygeno.2004.10.005 .

Morozova, O., Hirst, M. and Marra, M.A. 2009. "Applications of new sequencing technologies for transcriptome analysis." *Annual Review of Genomics and Human Genetics*, 10: 135–151. doi: 10.1146/annurev-genom-082908-145957.

Mafra, V., Martins, P.K. and Francisco, C.S. 2013. "*Candidatus Liberibacter americanus* induces significant reprogramming of the transcriptome of the susceptible citrus genotype." *BMC Genomics* 14 (1): 247 doi: https://doi.org/10.1186/1471-2164-14-247.

Martin, J., Bruno, V.M. and Fang, Z. 2010. "Rnnotator: an automated de novo transcriptome assembly pipeline from stranded RNA-Seq reads." *BMC Genomics* 11 (1): 663. doi: https://doi.org/10.1186/1471-2164-11-663.

Martzivanou, M. and Hampp, R. 2003. "Hyper-gravity effects on the Arabidopsis transcriptome. Physiology." *Plant* 118, 221–231. doi: https://doi.org/10.1034/j.1399-3054.2003.00092.x.

Miura, K., Ashikari, M. and Matsuoka, M. 2011. "The role of QTLs in the breeding of high-yielding rice." *Trends in Plant Science* 16, 319–326. doi: https://doi.org/10.1016/j.tplants.2011.02.009.

Muchero, W., Diop, N.N., Bhat, P.R., Fenton, R.D., Wanamaker, S., Pottor, M., Hearne, S., Cisse, N., Fatokun, C., Ehlers, J.D., Roberts, P.A. and Close, T.J. 2009. "A consensus genetic map of cowpea (*Vigna unguiculata* (L) Walp.) and synteny based on EST-derived SNPs. Proceedings of the National Academy of Sciences of the United States of America 106(43), 18159–18164. doi: https://doi.org/10.1071/FP1319 1.

Munroe, D.J. and Harris, T.J. 2010. "Third-generation sequencing fireworks at Marco Island." *Nature Biotechnology* 28, 426–428. doi: https://doi.org/10.1038/nbt0510-426.

Nair, P., Misra, A., Singh, A. and Shukla, A.K. 2013. "Differentially expressed genes during contrasting growth stages of Artemisia annua for artemisinin content." *PLoS ONE,* 8, e60375. doi: https://dx.doi.org/10.1093%2Fjxb%2Ferz166.

Nie, D.M., Ouyang, Y.D., Wang, X. and Zhou, W. 2013. "Genome-wide analysis of endosperm-specific genes in rice." *Gene* 530, 236– 247. doi: https://doi.org/10.1016/j.gene.2013.07.088.

Naithani, S., Preece, J., D'Eustachio, P., Gupta, P., Amarasinghe, V. and Dharmawardhana, P.D. 2017. "Plant Reactome: a resource for plant pathways and comparative analysis." *Nucleic Acids Research* 45:D1029–D1039. https://doi.org/10.1093/nar/gkw932.

Nagaraj, S.H., Gasser, R.B. and Ranganathan, S. 2006. "A hitchhiker's guide to expressed sequence tag (EST) analysis." *Briefings in Bioinformatics* 8 (1): 6–21. doi: https://doi.org/10.1093/bib/bbl015.

Nielsen, K.L., Hogh, A.L. and Emmersen, J. 2006. "DeepSAGE–digital transcriptomics with high sensitivity, simple experimental protocol and multiplexing of samples." *Nucleic Acids Research,* 34: e133.doi:10.1093/nar/gkl714.

Nagalakshmi, U., Wang, Z. and Waern, K. 2008. "The transcriptional landscape of the yeast genome defined by RNA sequencing." *Science* 320: 1344–1349. doi: https://doi.org/10.1126/science.1158441.

Oshlack, A., Robinson, M.D. and Young, M.D. 2010. "From RNA-seq reads to differential expression results." *Genome Biology* 11: 220. doi: https://doi.org/10.1186/gb-2010-11-12-220.

Palmieri, D.A., Novelli, V.M. and Bastianel, M. 2007. "Frequency and distribution of microsatellites from ESTs of citrus." *Genetics and Molecular Biology* 30 (3): 1009–1018. doi: http://dx.doi.org/10.1590/S1415-47572007000500029.

Parida, S.K., Anand Raj Kumar, K., Dalal, V., Singh, N.K. and Mohapatra, T. 2006. "Unigene derived microsatellite markers for the cereal genomes." *Theory of Applied Genetetic* 112, 808–817. doi: http://dx.doi.org/10.1007/s00122-005-0182-1.

Peleman, J.D. and Van der Voort, J.R. 2003. "Breeding by design." *Trends Plant Science* 8, 330-334. https://doi.org/10.1016/s1360-1385(03)00134-1.

Perez-de-Castro, M.A., Vilanova, S., Cañizares, J., Pascual, L., Blanca, J.M., Diez, M.J., Prohens, J. and Picó, B. 2012. "Application of genomic tools in plant breeding." *Current Genomics* 13 (3), 179–195. doi: https://doi.org/10.2174/138920212800543084.

Qiao, G., Wen, X., Yu, L. and Ji, X. 2012. "Identification of differentially expressed genes preferably related to drought response in pigeon pea (*Cajanus cajan*) inoculated by arbuscular mycorrhizae fungi (AMF)." *Acta Physiologiae Plantarum* 34(5), 1711–1721. doi: 10.1007/s11738-012-0966-2.

Raju, N.L., Gnanesh, B.N., Lekha, P., Jayashree, B., Pande, S., Hiremath, P.J., Byregowda, M., Singh, N.K. and Varshney, R.K. 2010. "The first set of EST resource for gene discovery and marker development in pigeonpea (*Cajanus cajan* L.)." *BMC Plant Biology* 10, 45. doi: https://doi.org/10.1186/1471-2229-10-45.

Ristaino, J.B. 2002. "Tracking historic migrations of the Irish potato famine pathogen, *Phytophthora infestans.*" *Microbes and Infection* 4, 1369–1377. doi: 10.1016/s1286-4579(02)00010-2.

Ritchie, M.E., Phipson, B., Wu, D., Hu, Y., Law, C.W. and Shi, W. 2015. "Limma powers differential expression analyses for RNA-sequencing and microarray studies." *Nucleic Acids Research* 43:e47.doi: https://doi.org/10.1093/nar/gkv007.

Reimand, J., Kull, M., Peterson, H., Hansen, J. and Vilo, J. 2007. "G:profiler—a web-based toolset for functional profiling of gene lists from large-scale experiments." *Nucleic Acids Research*, 35:W. doi: 10.1093/nar/gkm226.

Robinson, M.D., McCarthy, D.J. and Smyth, G.K. 2010. "EdgeR: a bioconductor package for differential expression analysis of digital gene expression data." *Bioinformatics* 26: 139–140. doi: 10.1093/bioinformatics/btp616.

Robertson, G., Schein, J. and Chiu, R. 2010. "De novo assembly and analysis of RNA-seq data." *Nature Methods*, 7 (11): 909–912.doi:10.1038/nmeth.1517.

Rudd, S. 2003. "Expressed sequence tags: alternative or complement to whole genome sequences?" *Trends in Plant Science*, 8 (7): 321–329. doi: 10.1016/S1360-1385(03) 00131-6.

Ruttink, T., Sterck, L., Rohde, A. and Bendixen, C. 2013. "Orthology guided assembly in highly heterozygous crops: Creating a reference transcriptome to uncover genetic diversity in *Lolium perenne.*" *Plant Biotechnology Journal*, 11, 605–617. doi: 10.1111/pbi.12051.

Redman, J.C., Haas, B.J., Tanimoto, G. and Town, C.D. 2004." Development and evaluation of an Arabidopsis whole genome Affymetrix probe array." *The Plant Journal* 38: 545–561. doi: 10.1111/j.1365-313X.2004.02061.x.

Rodrigues, C.M., de Souza, A.A. and Takita, M.A. 2013. "RNA-Seq analysis of Citrus reticulata in the early stages of *Xylella fastidiosa* infection reveals auxin related genes as a defense response." *BMC Genomics* 14 (1): 676. doi: https://doi.org/10.1186/1471-2164-14-676.

Salazar, M.M., Nascimento, L.C. and Camargo, E.L.O. 2013. "Xylem transcription profiles indicate potential metabolic responses for economically relevant characteristics of Eucalyptus species." *BMC Genomics* 14 (1): 201. doi: https://doi.org/10.1186/1471-2164-14-201.

Sanchez-Monge, E. 1993. "Introduction. In: Plant Breeding." Principles and prospects.M.D.Hayward, N.O.Bosemark, and I. Romagosa (eds.) Chapman and Hall, London, Glasgow, New York, Tokyo, Melbourne, Madras, pp. 3-5.

Sanger, F., Nicklen, S. and Coulson, A.R. 1977. "DNA sequencing with chain terminating inhibitors." Proceedings of the National Academy of Sciences U.S.A., 74: 5463–5467. doi: https://doi.org/10.1073/pnas.74.12.5463.

Sarkar, N.K., Kim, Y.K. and Grover, A. 2014. "Co expression network analysis associated with call of rice seedlings for encountering heat stress." *Plant Molecular Biology* 84, 125–143. doi: 10.1007/s11103-013-0123-3.

Sato, Y., Baltazar, A.A. and Nobukazu, N. 2011. "RiceXPro: a platform for monitoring gene expression in japonica rice grown under natural field conditions." *Nucleic Acids Research* 39 (1): D1141–D1148. doi: 10.1093/nar/gkq1085.

Schena, M., Shalon, D., Davis, R.W. and Brown, P.O. 1995. "Quantitative monitoring of gene expression patterns with a complementary DNA microarray." *Science*, 270 (5235): 467–470. doi: https://doi.org/10.1126/science.270.5235.467.

Schmid, M., Davison, T.S. and Henz, S.R. 2005. "A gene expression map of Arabidopsis thaliana development." *Nature Genetics* 37 (5): 501–506. doi: https://doi.org/10.1038/ng1543.

Schwacke, R., Ponce-Soto, G.Y., Krause, K., Bolger., A.M., Arsova, B. and Hallab, A. 2019. "MapMan4: a refined protein classification and annotation framework applicable to multi-omics data analysis." *Molecular Plant*. https://doi.org/10.1016/j.molp.2019.01.003.

Septiningsih, E.M., Pamplona, A.M., Sanchez, D.L. and Neeraja, C.N. 2009. "Development of submergence-tolerant rice cultivars: The Sub1 locus and beyond." *Annals of Botany,* 103, 151–160. doi: https://dx.doi.org/10.1093%2Faob%2Fmcn206.

Shankar, A., Singh, A., Kanwar, P. and Srivastava, A.K. 2013. "Gene expression analysis of rice seedling under potassium deprivation reveals major changes in metabolism and signaling components." *PLoS ONE,* 8, e70321. doi: https://doi.org/10.1371/journal.pone.0070321.

Sharma, R., Agarwal, P., Ray, S. and Deveshwar, P. 2012. "Expression dynamics of metabolic and regulatory components across stages of panicle and seed development in indica rice." *Funct. Integr. Genomics* 12, 229–248. doi: 10.1007/s10142-012-0274-3.

Shendure, J. and Ji, H. 2008. "Next-generation DNA sequencing." *Nature Biotechnology*, 26 (10): 1135–1145. doi: https://doi.org/10.1038/nbt1486.

Shiferaw, B., Smale, M., Braun, H.J., Duveiller, E., Reynolds, M. and Muricho, G. 2013. Crops that feed the world 10. "Past successes and future challenges to the role played by wheat in global food security." *Food Security* 5, 291–317. doi: 10.1007/s12571-013-0263-y.

Smedley, D., Haider, S., Durinck, S., Pandini, L., Provero, P. and Allen. J. 2015. "The BioMart community portal: an innovative alternative to large, centralized data repositories." *Nucleic Acids Research* 43:W589–W598. doi: https://doi.org/10.1093/nar/gkv350.

Soneson, C. and Delorenzi, M. 2013. "A comparison of methods for differential expression analysis of RNA-seq data." *BMC Bioinformatics* 14 (1): 91. doi: https://doi.org/10.1186/1471-2105-14-91.

Song, Y., Ma, K., Ci, D., Zhang, Z. and Zhang, D. 2013. "Sexual dimorphism floral microRNA profiling and target gene expression in andromonoecious poplar (*Populus tomentosa*)." *PLoS ONE,* 8, e62681. doi: https://doi.org/10.1371/journal.pone.0062681.

Surget-Groba, Y. and Montoya-Burgos, J.I. 2010. "Optimization of de novo transcriptome assembly from next-generation sequencing data." *Genome Research* 20 (10): 1432–1440. doi: https://doi.org/10.1101/gr.103846.109.

Swanson-Wagner, R., Briskine, R. and Schaefer, R. 2012. "Reshaping of the maize transcriptome by domestication." Proceedings of the National Academy of Sciences U.S.A., 109 (29): 11878–11883. doi: https://dx.doi.org/10.1073%2Fpnas.1201961109.

Sweetman, C., Wong, D.C., Ford, C.M. and Drew, D.P. 2012. "Transcriptome analysis at four developmental stages of grape berry (*Vitis vinifera* cv. *Shiraz*) provides insights into regulated and coordinated gene expression." *BMC Genomics* 13, 691. doi: https://doi.org/10.1186/1471-2164-13-691.

Tatum, L.A. 1971. "The Southern Corn Leaf Blight Epidemic." *Science* (80), 173, 39–39. doi: https://doi.org/10.1126/science.173.3991.39.

Tester, M. and Langridge, P. 2010. "Breeding technologies to increase crop production in a changing world." *Science* 327, 818-822. doi: https://doi.org/10.1126/science.118370 0.

Thudi, M., Li, Y., Jackson, S.A., May, G.D. and Varshney, R.K. 2012. "Current state-of-art of sequencing technologies for plant genomics research." *Briefings in Functional Genomics* 11(1), 3–11. doi: https://doi.org/10.1093/bfgp/elr045.

Trapnell, C., Pachter, L. and Salzberg, S.L. 2009. "TopHat: discovering splice junctions with RNA-Seq." *Bioinformatics* 25: 1105–1111. doi: https://doi.org/10.1093/ bioinformatics/btp120.

Trapnell, C., Williams, B.A. and Pertea, G. 2010. "Transcript assembly and quantification by RNA-Seq reveals unannotated transcripts and isoform switching during cell differentiation." *Nature Biotechnology* 28 (5): 511–515. doi: https://doi.org/10.1038/ nbt.1621.

Trick, M., Adamski, N.M. and Mugford, S.G. 2012. "Combining SNP discovery from next-generation sequencing data with bulked segregant analysis (BSA) to fine-map genes in polyploid wheat." *BMC Plant Biology* 12 (1): 14. doi: https://doi.org/10.1186/1471-2229-12-14.

Trick, M., Long, Y., Meng, J. and Bancroft, I. 2009. "Single nucleotide polymorphism (SNP) discovery in the polyploid Brassica napus using Solexa transcriptome sequencing." *Plant Biotechnology Journal* 7: 334–346. doi: https://doi.org/10.1111/j. 1467-7652.2008.00396.x.

UN World population projected to reach 9.8 billion in 2050, and 11.2 billion in 2100 Available online: https://www.un.org/development/desa/en/news/population/world-population-prospects-2017.html.

Varshney, R.K., Graner, A. and Sorrells, M.E. 2005. "Genic microsatellite markers in plants: Features and applications." *Trends Biotechnol.* 2005, 23, 48–55. doi: https://doi.org/10.1016/j.tibtech.2004.11.005.

Varshney, R.K. and Tuberosa, R. 2007a. "Genomics-assisted crop improvement vol. 1: Genomics approaches and platforms." *Springer;* New York. doi: https://doi.org/10. 1007/978-1-4419-0851-3_271.

Varshney, R.K.; Tuberosa, R. 2007b. "Genomics-assisted crop improvement vol. 2: Genomics applications in crops." *Springer*, New York. doi: https://doi.org/10.1007/ 978-1-4419-0851-3_271.

Väremo, L., Nielsen, J. and Nookaew, I. 2013. "Enriching the gene set analysis of genome-wide data by incorporating directionality of gene expression and combining statistical hypotheses and methods." *Nucleic Acids Research* 41:4378–4391. doi: https://doi.org/ 10.1093/nar/gkt111.

Varshney, R.K., Hiremath, P.J., Lekha, P., Kashiwagi, J., Balaji, J., Deokar, A.A., Vadez, V., Xiao, Y., Srinivasan, R., Gaur, P.M., Siddique, K.H.M., Town, C.D. and Hoisington, D.A. 2009a. "A comprehensive resource of drought- and salinity-responsive ESTs for gene discovery and marker development in chickpea (*Cicer arietinum* L.). *BMC Genomics* 10(1), 523. doi: https://doi.org/10.1186/1471-2164-10-523.

Varshney, R.K., Nayak, S.N., May, G.D. and Jackson, S.A. 2009b. "Next generation sequencing technologies and their implications for crop genetics and breeding." *Trends in Biotechnology* 27, 522–530. doi: https://doi.org/10.1071/FP13318.

Varshney, R.K., Glaszmann, J-C., Leung, H. and Ribaut, J.M. 2010. "More genomic resources for less-studied crops." *Trends in Biotechnology* 28: 452–460. doi: https://doi.org/10.1016/j.tibtech.2010.06.007.

Venu, R., Sreerekha, M. and Nobuta, K. 2011. "Deep sequencing reveals the complex and coordinated transcriptional regulation of genes related to grain quality in rice cultivars." *BMC Genomics* 12 (1): 190. doi: https://doi.org/10.1186/1471-2164-12-190.

Vidal, R.O., Mondego, J.M.C. and Pot, D. 2010. "A high-throughput data mining of SNPs in Coffea spp. suggests differential ESTs homeologous gene expression in the allotetraploid Coffea arabica." *Plant Physiology* (Bethesda), 154: 1053–1066. doi: https://doi.org/10.1104/pp.110.162438.

Vimolmangkang, S., Zheng, D., Han, Y. and Khan, M.A. 2014. "Transcriptome analysis of the exocarp of apple fruit identifies lightinduced genes involved in red color pigmentation." *Gene* 534, 78–87. doi: https://doi.org/10.1016/j.gene.2013.10.007.

Vettore, A.L., da Silva, F.R. and Kemper, E.L. 2003. "Analysis and functional annotation of an expressed sequence tag collection for tropical crop sugarcane." *Genome Research* 13 (12): 2725–2735. doi: https://dx.doi.org/10.1101%2Fgr.1532103.

Velculescu, V.E., Zhang, L., Vogelstein, B. and Kinzler, K.W. 1995. "Serial analysis of gene expression." *Science*-AAAS-Weekly Paper Edition, 270 (5235): 484–486. doi: https://doi.org/10.1126/science.270.5235.484.

Vodkin LO, Khanna A, Robin Shealy R, Steven J, Clough SJ, Gonzalez DO, Philip R, Gracia Zabala G, Thibaud-Nissen F, Sidarous M. 2004. "Microarrays for global expression constructed with a low redundancy set of 27500 sequenced cDNAs representing an array of developmental stages and physiological conditions of the soybean plant." *BMC Genomics* 5, 73. doi: https://doi.org/10.1186/1471-2164-5-73.

Von Braun, J., Rosegrant, M.W., Pandya-Lorch, R., Cohen, M.J., Cline, S.A., Brown, M.A. and Bos, M.S. 2005. "New risks and opportunities for food security scenario analyses for 2015 and 2050." *Food, Agriculture and Environment.* Discuss. Pap. 2005.

Voss-Fels, K.P. Stahl, A. and Hickey, L.T. 2019. "Q&A: Modern crop breeding for future food security." *BMC Biology* 17, 18. doi: https://doi.org/10.1186/s12915-019-0638-4.

Wang, F., Li, L., Li, H. and Liu, L. 2012. "Transcriptome analysis of rosette and folding leaves in Chinese cabbage using high-throughput RNA sequencing." *Genomics* 99, 299–307. doi: https://doi.org/10.1016/j.ygeno.2012.02.005.

Wang, L., Zhu, W., Fang, L. and Sun, X. 2014. "Genome-wide identification of WRKY family genes and their response to cold stress in *Vitis vinifera.*" *BMC Plant Biology* 14, 103.doi: https://doi.org/10.1186/1471-2229-14-103.

Wang, Z., Gerstein, M. and Snyder, M. 2009. "RNA-Seq: a revolutionary tool for transcriptomics." *Nature Reviews Genetics* 10 (1): 57–63. doi: https://doi.org/10.1038/nrg2484.

Wei, C.L., Ng, P. and Chiu, K.P. 2004. "5′ Long serial analysis of gene expression (LongSAGE) and 3′ LongSAGE for transcriptome characterization and genome annotation." Proceedings of the National Academy of Sciences U.S.A., 101: 11701–11706. doi: https://doi.org/10.1073/pnas.0403514101.

Walden, R. and Wingender, R. 1995. "Gene transfer and plant- regeneration techniques." TIBTECH 13: 324-331.

Wu, T., Qin, Z., Zhou, X., Feng, Z. and Du, Y. 2010. "Transcriptome profile analysis of floral sex determination in cucumber." *Journal of Plant Physiology* 167, 905–913. doi: https://doi.org/10.1016/j.jplph.2010.02.004.

Xu, Y., Lu, Y., Xie, C. and Gao, S. 2012. "Wholegenome strategies for marker assisted plant breeding." *Molecular. Breeding* 2, 29, 833–854. doi: http://dx.doi.org/10.1007/s11032-012-9699-6.

Yamamoto, N., Tsugane, T. and Watanabe, M. 2005. "Expressed sequence tags from the laboratory-grown miniature tomato (Lycopersicon esculentum) cultivar Micro-Tom and mining for single nucleotide polymorphisms and insertions/deletions in tomato cultivars." *Gene* 356: 127–134. doi: https://doi.org/10.1016/j.gene.2005.04.026.

Yang, Z., Zhang, T., Lang, T. and Li, G. 2013. "Transcriptome comparative profiling of barley eibi1 mutant reveals pleiotropic effects of HvABCG31 gene on cuticle biogenesis and stress responsive pathways." *International Journal of Molecular Sciences* 14, 20478–20491. doi: https://doi.org/10.3390/ijms141020478.

Yao, D., Zhang, X., Zhao, X. and Liu, C. 2011. "Transcriptome analysis reveals salt-stress-regulated biological processes and key pathways in roots of cotton (*Gossypium hirsutum* L.)." *Genomics* 98, 47– 55. doi: https://doi.org/10.1016/j.ygeno.2011.04.007.

Ye, J., Zhang, Y., Cui, H., Liu, J., Wu, Y. and Cheng Y. 2018. WEGO 2.0: a web tool for analyzing and plotting GO annotations, update. Nucleic Acids Research (46):W71–W75. https://doi.org/10.1093/nar/gky400.

Young N.D. and Bharti, A.K. 2012. "Genome-enabled insights into legume biology." *Annual Review of Plant Biology* 63, 283–305. doi:10.1146/annurevarplant-042110-103754.

Yu, Y., Guo, G., Lv, D. and Hu, Y. 2014. "Transcriptome analysis during seed germination of elite Chinese bread wheat cultivar Jimai 20." *BMC Plant Biology* 14, 20. doi: https://doi.org/10.1186/1471-2229-14-20.

Zan, Y., Ji, Y., Zhang, Y. and Yang, S. 2013. "Genome-wide identification, characterization and expression analysis of Populus leucine-rich repeat receptor-like protein kinase genes." *BMC Genomics* 14, 318. doi:10.1186/1471-2164-14-318.

Zhang, X., Rerksiri, W., Liu, A. and Zhou, X. 2013. "Transcriptome profile reveals heat response mechanism at molecular and metabolic levels in rice flag leaf." *Gene*, 530, 185–192. doi: https://dx.doi.org/10.1186%2Fs12864-018-4437-z.

Zerbino, D.R. and Birney, E. 2008. "Velvet: algorithms for de novo short read assembly using de Bruijn graphs." *Genome Research* 18 (5): 821–829. doi: https://doi.org/10.1101/gr.074492.107.

Zimmermann, P., Hirsch-Hoffmann, M., Hennig, L. and Gruissem, W. 2004. "GENEVESTIGATOR. Arabidopsis microarray database and analysis toolbox." *Plant Physiology* 136 (1): 2621–2632. doi: https://doi.org/10.1104/pp.104.046367.

About the Editors

Dr. Sarah Naiyer began her advanced training by completing a master's degree in Toxicology at the Jamia Hamdard University, followed by graduate research on the regulation of pre-rRNA processing and ribosome biogenesis in Entamoeba histolytica, at the Jawaharlal Nehru University in New Delhi, the premier research institution in India, with Prof. Sudha Bhattacharya. Dr. Naiyer was supported during her PhD studies by grants from the Council of Scientific and Industrial Research (CSIR) after having completed a qualifying examination in which she scored an All India Rank 31. Dr. Naiyer's graduate work resulted in several papers that were published in peer reviewed journals including Frontiers in Microbiology, BMC Genomics and JBC. Dr. Naiyer is currently working with Prof. Amy L. Kenter at The University of Illinois at Chicago, USA as a postdoctoral fellow, and has established an enhancer-enhancer interactome within the Igh locus, that contributes to the vast antibody repertoire formation (unpublished).

Dr. Khalid Hussain Bhat did his master's in biochemistry with a distinction from the University of Kashmir, India for which he received a university gold medal. After qualifying for a national fellowship for a research program, he moved to the Centre for DNA Fingerprinting and Diagnostics, Hyderabad, a premier basic research institute in India. Here he pursued his PhD in molecular immunology under the mentorship of Dr Sangita Mukhopadhyay. After completion of his dissertation, he joined Prof. Amy L. Kenter at The University of Illinois at Chicago, USA to investigate the mechanism of antibody repertoire generation. Dr Bhat has recently started his own laboratory at Sher-e-Kashmir University of Agricultural Sciences and Technology at Kashmir in India. Besides teaching undergraduate and graduate students, his research interests have expanded to plant microbes.

Index

H

I

M

N

P

R

S

T